# Spring Meeting Poster Session and Networking Reception 2018



Core Programming Area at the 2018 AIChE Spring Meeting and 14th Global Congress on Process Safety

Orlando, Florida, USA
22 – 25 April 2018

ISBN: 978-1-5108-6454-2

**Printed from e-media with permission by:**

Curran Associates, Inc.
57 Morehouse Lane
Red Hook, NY 12571

**Some format issues inherent in the e-media version may also appear in this print version.**



Printed by Curran Associates, Inc. (2018)

For permission requests, please contact AIChE
at the address below.

AIChE
120 Wall Street, FL 23
New York, NY 10005-4020

Phone: (800) 242-4363
Fax: (203) 775-5177

www.aiche.org

**Additional copies of this publication are available from:**

Curran Associates, Inc.
57 Morehouse Lane
Red Hook, NY 12571 USA
Phone: 845-758-0400
Fax: 845-758-2633
Email: curran@proceedings.com
Web: www.proceedings.com

# TABLE OF CONTENTS

# 514288 Low-Touch Machine Learning Is Fulfilling the Promise of APM (poster)

**Abstract:**

New methods and cutting-edge technologies are driving asset performance management (APM) well beyond historical capabilities, rapidly increasing its bottom-line value. Technologies such as cloud computing, data science and machine learning are now being integrated with automated methodologies directly into APM solutions.

This wave of integration firmly places advanced analytical techniques into the hands of operators and engineers with previously unimagined scale and ease of use. The incremental progress in APM over the last 20 years pales in comparison to what's now possible through digital transformation.

Low-touch machine learning is the key catalyst to scale APM's potential well beyond existing first principles-based solutions and "armies" of consultant engineers and data scientists. A widespread integration of machine learning in APM will mark a transition from estimated engineering and statistical models towards measuring actual patterns of asset behavior.

Manufacturing facilities staff can now readily extract value from decades of existing design and operations data to better manage and optimize asset performance. This "low-touch" machine learning method continuously embraces changes in asset behavior, empowering real-time APM value creation. Vetted and tested across diverse industries, scalable across multiple assets and powered by cloud and parallel computing, low-touch machine learning ushers in a new era of performance and optimization for all industries.

## 514063 Big Data Analytics Platforms: The Key to Enabling the Analytic Enterprise (poster)

**Abstract:**

Analytics and the Industrial Internet of Things are the new hot buzzwords these days, but analyzing data is nothing new, and manufacturers have been generating data off the line for decades. So why all the hype? The last few years have seen the evolution and advancement of several technologies that are converging to allow for unprecedented analysis and insights into business problems. But leveraging these takes more than just smart coders. Analytic insights need to be deployed into production and action has to be taken to truly realize the benefits. Find out why analytic platforms are the key to enabling the Analytic Enterprise.

## 514064 Industry 4.0 and Driving Better Organizational Decisions (poster)

**Abstract:**

Industry 4.0 is more than just about the Industrial Internet of Things, it is also about leveraging the Industrial Internet of People (IIOP). This talk will focus on using the tools now available to bring sensors, data, and analytics together to drive better organizational understanding and continuous improvement. Who are the key stake holders, and what do they need to improve to improve their contribution to the way a chemical plant is operated? How do you structure these projects to ensure successful deployments? This talk will focus on practical use cases and examples of Industry 4.0 in action.

# 514067 Early-Stage Digital Transformation, a Case Study (poster)

**Abstract:**

Enthought transforms the business of science with artificial intelligence. We are a world-class provider of scientific software solutions and digital transformation services to companies in a range of industries, from chemical manufacturing to oil and gas upstream research and production to aerospace, finance, and others. For more than 15 years Enthought has been building artificial intelligence and machine learning based solutions with science and engineering at the core.

In this talk, we present a case study of digital transformation at a polymer design and manufacturing company looking to improve their data management and increase extraction of the value hidden in their data. We will cover details of the on-going project, including steps taken to digitize previously manual, paper-based operations; changes in workflow and processes to support digitalization; and examples of the practical applications of machine learning techniques. In addition, we will discuss lessons learned, successes and pitfalls in the project. Positive transformations were not limited to the business only but also included new opportunities and personal transformation for employees and a positive change of culture at the company.

# 514070 Integrated Technical Computing for Process Engineering (poster)

**Abstract:**

Process engineering leverages data, domain knowledge, and modeling to design, scale, and continuously improve plants, from R&D to pilot to production. The workflow fundamentally involves data analysis and visualization, model development, and process optimization.

Data analysis starts with preprocessing of the dataset, to impute missing values, correct errors, and often impart structure. Descriptive analysis is used to characterize what has been observed, for each individual quantity, as well as compute correlations between quantities. Visualization is used to rapidly develop understanding regarding the relationships between predictors and responses.

Insights gained from data analysis are subsequently leveraged to create a model, to describe the responses as mathematical functions of the predictors. In simple cases, first principles understanding is confirmed, and a governing equation can be written. In cases of intermediate complexity, the basic mechanisms for observed behavior are described by a theoretical framework, and parameter estimation or process identification methods can be used. Complex cases are typified by a multiplicity of superposed physical effects and absence of an economic functional form â€"in these cases, machine learning techniques are required to infer input-output relationships.

Once the model has been created, predictions can be made about the responses that would be recorded if an experiment based on the predictor values was conducted. Since the model enables the parameter space to be explored continuously, optimization methods can be employed to find the combination of predictors that minimizes a specified function of the responses. The optimal conditions so determined inform the selection of the plant operating setpoint. Sensitivity studies are conducted to define the plant operating region around the setpoint, and thus specify the control system requirements.

There are numerous software tools available which address select subsets of the process engineering workflow. MATLAB is one of very few that cover it end-to-end, and is distinguished in that it also offers utilities for distributed computing, big data processing, automatic code generation, and real-time interfacing with other software tools â€"including process simulators.

In this talk we describe MATLAB functionality relevant to each stage of the process engineering workflow, and provide application examples and graphics to bring the concepts to life. Specific topics will include data analysis (tall objects, logical indexing, statistical plotting), model development (principal components, system identification, neural networks), and optimization (linear programming, pattern search, and genetic algorithms).

## 514071 Tony Starck Needs Jarvis to Run Iron Man. What's the Future in AI for Continuous Process Industries? (Poster)

**Abstract:**

Why bigdata is such a game changer? We have run plants with automation and scada levels for 30 years, and we encounter a lot of data. We visualize, we set, we adjust, we regulate. With more and more advanced technologies. We make a lot of decisions, impacting the performance of our plants. The bigdata era creates another level of opportunity. By crossing all kinds of information, we open the gate of a new worldâ€¦ of complexity. We understand that there is a gold mine awaiting our discovery, but our biological brain, the main technology we use to dig into it, is now outdated. Our brains will not crunch the bigdata! We need new technology : artificial intelligence.

Today, the most advanced continuous process plants are about to become autonomous. Transforming the decisions of process optimization to dedicated AI. This is coming fast for all of us. What does already exist? How do we implement it for what profit? What can we expect for the future? How can this benefit us?

# 514072 Approaches to Multiblock Modeling and Dimension Reduction for Combining Disparate Data Types (poster)

**Abstract:**

Continuous manufacturing, which is being strongly promoted by the FDA, is an innovative approach to pharmaceutical processing that utilizes information from multiple sensors throughout the entire process. Due to the numerous factors that need to be monitored and controlled at the same time, a multi-step approach is necessary for optimizing individual unit operations as well as the overall process. In this presentation we will show some approaches to combining process and spectroscopic data, which can be applied to process monitoring for continuous manufacturing as well as for batch processes.

# 513463 Manufacturing Analytics - the Journey so Far

**Abstract:**

The goal of Enterprise Manufacturing Intelligence (EMI) is to make these high value relevant data available in real time to all levels of production management (Operations to Plant and Business Management). Therefore, plants can take advantage of this information to make improvements in raw material usage, asset utilization by reducing unnecessary process wait times and customer satisfaction through improved product quality by reducing product and process variation. This applies to both batch and continuous processes.

At The Dow Chemical Company, we have been actively pursuing EMI since 2012. Since that time we have done 70+ EMI implementations in a variety of environments. This poster talk will discuss the different audiences / kinds of problems our installations have sought to address. We will also discuss some challenges (both technical and cultural) and how we overcame them. Finally we will discuss how these experiences have shaped our view of the future for EMI at Dow.

# 516856 Applying Predictive Analytics to Develop Actionable Demand Forecasts


**Abstract:**

The Dow Chemical Company is a leading chemical manufacturer in the world. Dow manufactures plastics, chemicals, and agriculture products. With a presence in over 150 countries, it employs over 50,000 people worldwide. Dow serves to customers in different market segments, with a wide variety of products made for each one. The magnitude of Dow makes demand forecasting a challenge. Dow is using hierarchical statistical demand forecasting techniques to accurately forecast demand for the key products and customers. This poster is about the data and predictive analytics methods and tools used to develop the actionable demand forecasts.

## 506402 Measurements and Calculations of Asphaltene DepositionÂ

**Abstract:**

Asphaltene precipitation and deposition have been recognized as curtailing and vital problems in oil and gas fields. Asphaltene precipitation and deposition affect production, transmission and processing facilities in different and tremendous ways. These problems lead to system interruptions and have become a crucial part of flow assurance studies. One way to mitigate these problems is to keep the production system free of wax and asphaltenes.

Different approaches have been implemented in the oil and gas industries to predict asphaltene precipitation or deposition conditions. Predicting the asphaltene onset pressure (AOP) requires the presence of pressurized down-hole samples with very low or no oilâ€“based mud contamination, which is not an easy task.

Several models have been introduced in the oil industry; some of them are considered costly and time consuming since they require experimental work to come up with an accurate prediction.

The asphaltene precipitation conditions in Kuwait oil fields from both South East Kuwait (SEK) and West Kuwait (WK) were studied extensively to develop a fast and low-cost model with acceptable accuracy.

This study has three phases. The first phase is a literature review regarding the available approaches to predict the AOP. Later, a new model is introduced to overcome the current industrial obstacles. The new model, which relies on basic reservoir data such as the temperature, pressure, bubble point and a SARA analysis, was tested using lab measurements.

In the final stage, PIPESIM software was used to predict the AOP for eighteen Kuwaiti oil wells. The accuracy of the proposed model as well as the PIPESIM predictions indicated that both approaches resulted in some error compared with the measured values of the AOP. The model developed in this study has an average absolute error of 12% compared to 7.50% by PIPESIM. However, the proposed model is easy to be used and does not need commercial license.

## 512068 Research on Solution of Suspending Seabed Pipeline Based on Bionic Protective Technology

**Abstract:**

The paper intends to propose an immediate and also permanent solution to suspending seabed pipeline, which is a fatal factor of the oil spill accident, caused by scour of currents, instability of the seabed, vibration of the riser and others developed by the present authors. Based on the marine dynamics and marine bionics, the mechanism and effect of scour is introduced and a new protective technique is explored. As core of technique, composite material fronts mats are secured to the seabed using anchors and the buoyant fronts reduce the current velocity and sediment transportation to protect the pipeline rely on viscous drag, causing sediment to fall into the fronds to back fill scour pits and to form a permanent reinforced sandbank. Compared with other traditional methods, which seldom prohibit second scour, the efficiency and feasibility of this technique could be guaranteed. In order to demonstrate the advancement, a set of model experiments are provided to show the status of deposition with time. Three cases of application in risers of an oilfield company indicate that the protective system with the sediment deposition has meaningful influence on elimination and rectification scour. With the characters of "one time", "one cost" installation, low installed cost, environmental friendly material and cheap maintaining, this protective technique has benefit to provide a permanent, self-maintaining solution to scour, improve of the integrity of the suspending seabed pipeline ,finally to control oil spill accident.

# 512539 A Heat-Recirculating Combustor Using SiC Porous Foam for Thermophotovoltaic Power Devices

**Abstract:**

Efficient and reliable portable power devices have been investigated due to the increasing demand of electronic devices. Because of the advantage of the high energy density, power systems using the combustion of hydrocarbon fuels are considered as one alternative to replace secondary batteries. A thermophotovoltaic (TPV) power system, which is one of the combustion-driven power systems, converts thermal energy to electricity through photons. The TPV power systems consist of simple structure and have high efficiency without friction loss by moving parts. To enhance the efficiency of the TPV systems, several design factors are important: stable burning in the combustor as a heat source and high enough temperature of the emitter as a radiation source. In this study, to improve the performance of the combustor that was developed for TPV power devices in a previous study, silicon carbide (SiC) porous foam is inserted into a combustion chamber. Porous burners have been considered as one possible technology for achieving the high-efficiency by increasing residence time for complete combustion and enhancing internal heat recirculation. To evaluate the combustor performance, combustion stability limits and temperature distribution along the emitter wall surface are measured. Compared with the case without SiC porous foam in the combustion chamber, the flame stability limits are extended with SiC porous foam, implying a widened operating range of the combustor. The temperature of the emitter wall surface is locally enhanced where SiC porous foam is installed, and the average temperature of the emitter wall also increases when SiC porous foam is used.

## 513034 Combustion Properties of Outwardly-Propagating Spherical CH4-NH3/Air Flames

**Abstract:**

Fundamental combustion characteristics of methane ($CH_4$)-ammonia ($NH_3$) blends are studied to evaluate the potential as a low-carbon fuel since their utilization can be considered as an interim technology for responding to global warming. The propagation of spark-ignited spherical laminar premixed $CH_4$-$NH_3$/air flames at normal temperature and pressure is experimentally investigated and also computationally predicted using a detailed kinetic mechanism. The fundamental unstretched laminar burning velocities and Markstein numbers are determined and the associated flame structure is investigated. Results show that the laminar burning velocities are reduced with $NH_3$ substitution, while their tendency with varying fuel-equivalence ratio is similar to the pure $CH_4$/air flames. Predicted flame structure exhibits that nitrogen oxide ($NO_x$) formation is enhanced with $NH_3$ substitution, but at fuel-rich conditions the increased amount of $NO_x$ emissions is much lower than that under fuel-lean conditions. Markstein numbers of $CH_4$-$NH_3$/air flames are similar to the pure $CH_4$/air flames, implying that the flamefront instabilities are not significantly affected by $NH_3$ substitution. Measured and predicted tendencies of the laminar burning velocities and Markstein numbers for various conditions show encouraging agreement, but quantitative discrepancies among the measurements and predictions at some conditions merit additional consideration.

## 511675 Implementation of Distributed Co-Simulation for Robust Design of Urban Energy Systems

**Abstract:**

Simulation can support an optimal design and operation strategy of multi-energy systems. One approach being co-simulation allows the use of dedicated software and tools for each subsystem.

Except for tools offering Functional Mock-up Interface (FMI) functionalities, there is no common standardized interface that can be used for the co-simulation of such multi-energy systems. Rather than trying to create new standards, the goal here is to create a robust and adaptable co-simulation platform, acting as a easily linkable backbone.

This presentation is the opportunity to show how we implemented a distributed multi-platform co-simulation using containerization concept and an orchestrator to link tools such as TRNSYS, EnergyPlus, Modelica, Matlab and Python based tools, with or without FMI.

Currently a lack of robust software implementation skills can lead to misused, unused or unmaintainable academic tools. The presented platform will enable non computer specialists to exploit the possibilities of co-simulation techniques while respecting standards for tool development that will make their contributions reusable by others and easily maintainable.

We will present the application of the co-simulation platform on a use case based on the city of Vevey (Switzerland) combining a low temperature district heating and cooling network with decentralized heat pumps, an electricity network with power to gas technologies and a gas powered combined heat and power production plant supplying a high temperature district heating network and compare the use of the distributed co-simulation platform with that of a classical monolithic simulation using Modelica.

This method is applied for the implementation of distributed multi-platform co-simulation for hybrid urban energy systems in the IntegrCiTy project funded by the ERA-NET Cofund Smart Cities and Communities.

## 512947 The Feasibility of Replacing Electric Drives with Wind Energy in the Swro System

**Abstract:**

The feasibility of substituting electricity in the Sea Water Reverse Osmosis (SWRO) process with wind energy is considered in this article.

The SWRO process is divided into very high and medium pressure areas. The medium pressure areas are listed below;

1. Sea water cleaning and conditioning section
2. High purity water conditioning
3. Product water export

The hydraulic system for the medium pressure areas consists of pipe networks that supply Pelton or turbine drives (P-drives) for pumps. The energy input to the hydraulic network is from wind turbine driven pumps.

The very high pressure area of the SWRO unit is a once through system. A wind turbine-pump pair capable of producing the required head is considered.

# 512962 Bio-Drying: A Sustainable Approach for Enhancing the Pre-Treatment Process and Thermal Properties of Green Waste

**Abstract:**

### Abstract

The potential for thermal recovery of waste is gaining impetus among researchers and industries across the globe especially in many developed countries. However, in processing waste for energy recovery process, the type and nature of input waste materials, particularly those with high moisture content, have a significant impact in determining the quality, environmental profile of the waste technology as well as the thermal properties of the final product. Bio-drying, as a waste to energy conversion technology, tends to reduce moisture content of waste while maintaining the energy content of the processed waste. The current study investigates the effect of input materials (plastics, sorted waste papers, pruning waste and food waste) on the energy content of waste material with subsequent implication on biogenic content of the material by bio-drying process. Preliminary results established a positive correlation between biogenic and energy content of the input materials with some variations observed between food and paper waste. High biogenic content was reported for plastic waste whereas paper waste had the highest non-biogenic content of 9%. Further analysis after the bio-drying process show that, depending on the proportion of the different waste materials, the biogenic and energy content of the final material differed indicating the significant of the amount of some materials in the waste mix with respect to others. Among the waste components plastic was identified as the major contributor to high biogenic and energy content, whereas paper was considered as the lowest contributor. Additionally, high proportion of food waste in the waste mix indicated a slight difference in biogenic and energy content; on the other hand, the same proportion of paper in the waste mix showed similar biogenic content with some variation of energy content observed. This research revealed that the combination of green waste of high moisture content with other waste materials by bio-drying process would offer alternative waste management approach and help mitigate environmental pollution from the disposal of biodegradable waste.

**Keywords:** Bio-drying; Biogenic content; Energy content;

# 510750 Towards Standardisation of Process Intensification in Solids-Handling Processes – the Ibd Platform

**Abstract:**

Process Intensification (PI) is broadly defined as *any chemical* (or process) *engineering development that leads to substantially smaller, cleaner, safer and more energy efficient technologies* and is achieved by significant enhancement of transport rates. To date, the majority of achievements in process intensification have been in liquid and/or gaseous phase applications.

Intensification of processes in which solids are an intrinsic component is significantly more difficult, with fouling an inherent problem which is contradictory to one of the key PI aims of equipment: miniaturisation. However, there are various examples of successful PI technologies across the broad spectrum of solids handling, including but not limited to the Taylor-Couette reactor, the twin-screw granulator and the toroidal fluidised bed.

Intensified by Design (IbD) is a large, multi-partner project funded by the European Union Horizon 2020 programme (SPIRE-08-2015, Grant No. 680565) which brings together a team of experts in PI, fouling, process analytics, process modelling and process control from across Europe. The main project output is the IbD Platform, a holistic, computer-based methodology for undertaking process intensification in solids handling processes. Features of the IbD Platform include:

- A comprehensive database of existing PI technologies and a knowledge-based selection methodology for initial shortlisting of options.
- A number of "*built-in-modules*" giving comprehensive design procedures for technologies from the PI database.
- A PI-based TRIZ methodology to guide new process innovations.
- Strategies for fouling prevention and remediation.
- Strategies and tools for advanced monitoring and control.

The IbD Platform and methodologies are validated against 6 industrial case studies from a range of sectors: mineral processing, ceramics processing, pharmaceuticals and metallic nano-particle production.

# 513010 Process Design and Development of Offshore NGL (Natural Gas Liquid) Recovery Processes

**Abstract:**

**Process Design and Development of Offshore NGL (Natural Gas Liquid) Recovery Processes**

Sekwang Yoon[1], Jin-Kuk Kim[1], Chi-Kyun Park[2], Byung-Don Lee[2]

[1] Department of Chemical Engineering, Hanyang University, Wangsimni-ro 222,

Seongdong-gu, Seoul, 04763, Republic of Korea;

[2] Institute of Chemical & Environmental Process, Jeonjin Entech Co., Ltd, Gadal 1-Ro 38, Gangseo-Gu, Busan, 46729, Republic of Korea;

This work focuses on process design and development for the recovery of ethane (or propane) from natural gas liquids (NGL) to be applicable in the offshore environment. In this work, the NGL recovery is mainly made through the distillation at sub-ambient conditions, and low temperature condition is created by the use of a turbo-expander as well as J-T valves, which are interconnected with a demethanizer column.

Considerable efforts have been made in process design to improve energy efficiency of NGL recovery process without compromising product specifications. First, heat integration techniques have been applied to maximize heat recovery existed within the process. Second, attempts have been made to fully realize any potentials for power recovery. Third, column analysis has been carried out to identify any inefficient features in column design, and column is designed in an energy-efficient manner by introducing a few side-reboilers and selecting optimal flowrate and operating conditions for feed streams. Finally, optimization method has been rigorously applied to determine the most appropriate configurational features (i.e. interconnections between a column, heat exchangers and a turbo-expander.) as well as optimal operating conditions.

Offshore characteristics have been incorporated in the design of demethanizer column in this study, in which techno-economic impacts caused from ship motions are systematically considered. This work will also address control and operability issues, in the context of maintaining design feasibility of NGL recovery process, and achieving reliable and safe operation in industrial practices.

*This research was supported from the Technology Innovation

(10042424, Floating Production Platform Topside Systems and Equipment Development) funded By the Ministry of Trade, Industry & Energy (MI, Korea).

Korea Institute for Advancement of Technology(KIAT) through the Encouragement Program for The Industries of Economic Cooperation Region)(R0003996)

## 513304 *Thermodynamic Data Requirements for Biodiesel Process Development*

**Abstract:**

Biodiesel has been the drop in solution for the renewable energy production particularly from various waste sources. Many new technologies are emerging with data demands for efficient scale-up and process design. The presentation will highlight the data requirements, gaps in the thermophysical property data, and consolidation of experimental and estimation methods for effective process development. A case study on how new calorimetric studies for heat of mixing measurement and Aspen flowsheet simulator were utilized for development of a 100,000 gallon/year continuous mobile mini biodiesel plant.

## 513337 Enzymatic Reaction Kinetics for Biodiesel Process Development and Scale-up

**Abstract:**

Biodiesel production from waste resources has been widely practiced as a possible solution to the increasing demand for renewable energy. Many new technologies for this production have been developed and require experimental data for efficient process design and scale-up. One such technology is the use of immobilized enzymes for biodiesel production. This presentation highlights the required kinetic data for effective process development of enzymatic biodiesel production using process simulation software. Isothermal calorimetric experiments and an Aspen process simulator were utilized for the development of a 100,000 gallon/year continuous miniature biodiesel plant.

# 513405 Direct Process for Methylchlorosilanes: Chemistry and Technology


**Abstract:**

Dow Performance Silicones (DPS) is the world leader in the direct process technology for the production of methylchlorosilanes (MCS, $Me_xSiCl_y$). MCS is the intermediate product to produce Poly Dimethyl Siloxanes (PDMS) that is the core base material to produce several thousand final silicone polymers. Silicones are synthetic polymers that combine the characteristics of plastics and glass with mixed properties of organic and inorganic polymers. MCS are produced *via* the Direct Process reaction of silicon metal with methyl chloride gas in the presence of copper catalysts and other metal promoters such as Sn, P, and Zn. Direct process is a unique solid-gas heterogeneous catalytic reaction that is more efficient with a bulk metal catalyst than a supported metal catalyst. The direct process catalysis involves generation of a consumable reactive site that repeats the cycle *via* regenerating the catalyst. This is unlike the *active site* regeneration in traditional heterogeneous catalytic processes. Reaction deactivates by the accumulation of impurity metals on the surface and methyl group decomposition that cause diffusion and kinetic hindrance as well as surface coking. This poster intends to highlight the key catalysis mechanism and technology aspects of this major commercial process and challenges and opportunities that exist in process improvement.

## 514278 Development and Application of Simulation Models for Improvement of Highly Non-Ideal and Coupled Distillation Systems

**Abstract:**

An organic acid-water mixture is separated using an entrainer which forms an azeotrope with water. The mixture also contains smaller amounts of other components which are separated in downstream distillation columns. The system is highly non-ideal with significant recycle between these columns.

After the startup, desired separation was not achieved resulting in the loss of organic acid. A simulation model was developed to represent this system consisting of 5 coupled columns. Several case studies were run to understand the impact of impurities and other operating parameters. These case studies helped to stabilize and improve the performance of the columns.

# 515353 Effect of Electrostatic Field on Asphaltene Aggregation and Deposition Under Dynamic Conditions

**Abstract:**

**Effect of Electrostatic Field on Asphaltene Aggregation and Deposition Under Dynamic Conditions**

Vahid Azari, Erfan Abolghasemi, Shahab Ayatollahi and Vahid Taghikhani

Department of Chemical and Petroleum Engineering, Sharif University of Technology, Tehran, Iran.

**Abstract**

Asphaltene deposition during oil production is a leading cause in decreasing the production rate. Once occurred in an oil well, it could drastically increase the cost of remediation, workover and production. Therefore, preventative actions to control the asphaltene deposition and remediation of such heavy deposits which leads inevitably to reduction or even cessation in oil production is a main challenge. It has been shown that depending upon the nature of asphaltene molecules and the surrounding solvents, the asphaltene molecules can significantly react against electrostatic field. In other words, by applying electrostatic field precipitation, aggregation and deposition of asphaltene molecules can be controlled. To continue the previous research, in this work, the effect of electrostatic field under dynamic conditions was investigated on different crude oil samples. To do so, a homogenized mixture of crude oil and solvent was pumped into an open channel at various crude oil/asphaltene slurry flow rates using a peristaltic pump. Simultaneously, two parallel side walls of the open channel were connected to the positive and negative terminals of a DC power supply to turn them into the anode and the cathode electrodes in the electric circuit. The results showed that the asphaltene molecules, studied in this work, had tendency to be deposited on the negative electrode. Therefore, the residual surface electric charge on the colloidal molecules of asphaltene in crude oil was confirmed to be positive. Also, the effects of electric voltage, solvent/crude oil ratio and the flow rate of asphaltene/crude oil slurry in the open channel were studied on the amount of asphaltene deposition. It was shown that the electrostatic field can highly affect the aggregation rate and the deposition of asphaltene particles under flow conditions. Increasing the amount electric voltage can increase the aggregation rate of asphaltene molecules as well as and the average size of aggregates. This, in turn, can affect the rate of deposition of asphaltene molecules on the negative electrode. Also, it was concluded that an optimal crude oil flow rate was needed to be used in order to control and minimize the amount of asphaltene deposition. The results showed that the technique, proposed in this study, can be used to manage asphaltene aggregation and deposition as a preventative method. Lastly, the potential of practical application of this technique on a field scale was also discussed.

# 511587 Investigation of Bentonite Swelling Kinetics Using Particle Size Analysis

**Abstract:**

Structural fluids contain more than one phase within its solution, a liquid continuous phase with solid dispersed phase. Such fluids are widely used in most of manufacturing processes: food, pharmaceuticals, paints, plastic and solvents. In drilling of oil and gas wells, bentonite is one of the main component in formulating the drilling fluids. Bentonite dispersion in water forms a viscous structural fluid. Dispersion depends on variable factors such as particles size, shape, concentration, ion type and attraction force. Viscosity measurement for these structural fluids is critical and time-consuming process. In this paper, a low swelling Ca-bentonite is activated using soda ash combined with thermal treatment. The bentonite was successfully activated to increase and Na/Ca and hence improve the swelling capacity. Swelling of bentonite was investigated using particles size analysis. The particles size distribution distortion with time and temperature was analyzed using simple kinetic chemical reactions models. The results of particles size distribution were in agreement with viscosity measurement. The swelling kinetics of the treated bentonite fits a second â€"order model. The activation energy and the arte constants were obtained.

**Acknowledgement**

The authors would like to acknowledge the support of Qatar University and KFUPM for this research.

# 512852 Improved Oil Recovery with Viscoelastic Surfactant: Screening and Effectiveness of Formulations

**Abstract:**

Middle East carbonate reservoirs are usually characterized as high temperature and high saline (HTHS) reservoirs. In chemical enhanced oil recovery (cEOR), interfacial tension (IFT) and capillary number are the key parameters that alter the reservoir environment. Viscoelastic surfactant (VES) is introduced due to its dual capability of lowering the IFT and improving the sweep efficiency of the reservoir. In this work, the main objective is to screen out a novel VES for cEOR applications in HTHS carbonate reservoirs.

Novel VES solutions of 0.1 to 1 wt% concentrations were prepared with synthetic sea water and stirred until they attain equilibrium. Densities and IFTâ€™s of VES solutions and oil at different temperatures (30°C, 50°C and 80°C) were determined using density meter and spinning drop tensiometer (Kruss). Equilibrated solutions were used with heavy crude oil of 17°API to determine the IFT. Both short term and long term thermal stability of VES was evaluated using an approach based on FTIR, NMR and TGA analyses. Aging of VES was performed for 30 days at two high temperatures of 90°C and 120°C.

IFT results showed considerable reduction in IFT with novel VES and heavy oil upto $10^{-1}$ mN/m range. IFT decreased initially and then increased with increase in VES concentrations. Non-linear trend of IFT was obtained. IFT reduction usually depends on the concentration of solutions used. Results also indicate that ultralow IFT was obtained at 50°C. Critical micelle measurements were performed and 3wt% was found as CMC. VES was evaluated for short term thermal stability using TGA analysis and found to be stable upto 250°C. FTIR, $^{1}$H NMR and $^{13}$C NMR before and after aging the VES samples were compared and structural changes were observed in the aliphatic region. The peaks were shifted away due to the presence of C-O bond in the sample. IFT for aged samples was determined and almost two folds of increase in IFT was observed. Depending upon the results of this investigation at high temperatures from FTIR, $^{1}$H NMR and $^{13}$C NMR and IFT, this novel VES exhibit great potential to be used in cEOR applications when we have HTHS reservoirs.

# 513864 Protecting Environment and Assuring Smooth Energy Transfer Using Ionic Liquids

**Abstract:**

Gas hydrates are ice-liked crystalline compounds that are formed when small gas molecules get trapped within the water molecules under high pressure and low temperature conditions in oil and gas transmission lines. The formation of these hydrates is a major threat to oil and gas industry as they have tendency to agglomerate and completely block the oil and gas transmission lines, which may lead to an explosion or cause unwanted operations shut down. Therefore, annually industry spends around 1 billion US dollars on hydrate prevention procedures which includes extensive use of chemical inhibitors.

These chemical inhibitors are generally classified as thermodynamic hydrate inhibitors (THI) and kinetic hydrate inhibitors (KHI). The thermodynamic hydrate inhibitors function by shifting hydrate dissociation temperature to lower values and kinetic hydrate inhibitors function by delaying the hydrate formation time. The commercial THI like Methanol and Mono-ethylene glycol (MEG) perform well, but these inhibitors are required in large quantities (> 30 wt%) and cannot be easily disposed of into the environment. Therefore, there is a strong industrial need to design inhibitors that are environmentally friendly and are required in low dosage.

Ionic liquids (ILs) well known as ionic fluids are type of an organic salts that have low melting points and tendency to stay in a liquid form at low or ambient temperature. Ionic liquids are extensively being used in different chemical processes due to their negligible vapour pressure and low viscosity. Recently, ionic liquid have been recognized as the dual functional inhibitors as they have tendency to perform as kinetic hydrate inhibitor and thermodynamic hydrate inhibitor simultaneously.

In this experimental based work the thermodynamic inhibition (TI) and kinetic inhibition (KI) effect of ionic liquids (ILs) 1-Methyl-1-Propyl-pyrrolidinium Chloride [PM-Py][Cl] and 1-Methyl-1-Propyl-pyrrolidinium Triflate [PM-Py][Triflate] have been investigated on a methane rich gas mixture at different concentrations and pressure ranges (40-120 bars). The effect of addition of synergents with ionic liquids has been also studied and the experimental results have been compared with the commercial thermodynamic inhibitor methanol and literature data. All the experimental work has been conducted using PSL systemtecknik rocking cell assembly.

**Acknowledgement**

*This work was made possible by NPRP grant # 6-330-2-140 and GSRA # 2-1-0603-14012 from the Qatar National Research Fund (a member of Qatar Foundation). The statements made herein are solely the responsibility of the authors.*

# 512893 Flow Analysis of an Atypical Cylindrical Cyclone Separator with a Large Ratio of Length to Diameter

**Abstract:**

This study presents a flow analysis on an atypical cylindrical cyclone separator with a large ratio of length to diameter. Due to the distribution of the flow field determines the merits of separation performance, the aim is to optimize the flow. The Phase Doppler Particle Analyzer (PDPA) is used to measure the gas flow, and the numerical simulation is carried out by using the Reynolds Average Navier-Stokes (RANS) equation, the Reynolds stress model (RSM). Since the model is validated by good agreement between the numerical results and experimental data, and then the tangential velocity is analyzed from the transient flow field and dynamic performance.

The results showed that the tangential velocity of the transient flow field existed a non-axisymmetric phenomenon in the cross section, which was mainly presented as asymmetric distribution of the contours. The zero location of tangential velocity value was not coincide with the geometric center, where the tangential velocity was larger near the zero location. In addition, the dynamic character of tangential velocity was high-speed pulsating, with quasi-periodic characteristic. The structure or operation optimization could change the pulsation frequency and reduce the industrial vibration, which is highly significant.

Meanwhile, a parameter of RMS was used to identify the degree of turbulence velocity deviating from mean velocity. The RMS data showed that the flow instability gradually increased with the fluid flowing into the cyclone separator from the inlet section, until reached a peak value at the position about 2.0 times the length to diameter. And then the flow instability gradually decreased, with the fluid energy dissipated.

## 512855 Laboratory Investigation of Viscoelastic Surfactant for Chemical EOR in High Salinity and High Temperature Carbonate Reservoirs

**Abstract:**

Viscoelastic surfactants are gaining much attention in the recent years due to their structure. Acquiring dual capabilities of a polymer as well as a surfactant in one composition provides considerable value to chemical EOR. Middle East reservoirs are characterized as high salinity and high temperature that make carbonate reservoirs in this region even more challenging. In this work, a systematic experimental study is carried out to test the applicability of novel viscoelastic surfactant (VES-AGA 97) in enhancing the oil recovery.

Rigorous investigation of VES was performed to determine its potential in improving oil recovery. Screening of VES was carried out using five major evaluation techniques. Thermal stability of VES was determined using TGA and FTIR methods. Interfacial tension (IFT) measurements were recorded using spinning drop tensiometer. $^{1}H$ NMR and $^{13}C$ NMR were measured with long-term thermal stability tests performed for 30 days in anaerobic conditions at two different high temperatures. Static adsorption was measured by shaking the crushed powder samples with VES for 24 hours and measuring the changes in concentration of VES. Dynamic adsorption is estimated by performing core flooding experiments at 90°C.

Experimental results with TGA demonstrated that VES was thermally stable and showed high resistance to temperatures up to 250°C. FTIR results indicated good thermal stability with strong bonding between C-H and C=O. Considerable reduction in IFT with VES up to $10^{-1}$ mN/m range was observed. Effect of concentration, time and temperature on IFT was evaluated. It was also noted that ultralow IFT was obtained at 50°C and 80°C. $^{1}H$ NMR showed stability in structure while $^{13}C$ NMR results showed minor structural changes in the aliphatic region. Static adsorption results gave less adsorption when we had high temperatures compared to low temperatures. Langmuir and BET adsorption isotherm models were utilized to determine maximum adsorption rate. Comparison between both models and experimental results exhibited perfect match. Dynamic adsorption was evaluated using concentration analysis of effluents collected at the downstream from core flooding. Difference between the concentrations of VES at inlet and outlet was measured using total organic carbon analysis. Additional oil recovery of 8% was determined when recovery factor is plotted against pore volumes.

# 513411 New Air Assisted Flare Design to Handle Low Flow Rates of Waste Gases

**Abstract:**

New Air Assisted Flare Design to handle Low Flowrate of Waste Gases

Hayder A. Alhameedi [a], Joseph D. Smith [b]

[a] Department of Chemical and Biochemical Engineering, Missouri University of Science and Technology, Rolla, MO 65409-1230. USA, +1 (573) 612-8911, haa6hb@mst.edu

[b] Department of Chemical and Biochemical Engineering, Missouri University of Science and Technology, Rolla, MO 65409-1230. USA, +1 (573) 341-4294, smithjose@mst.edu

Abstract

Flares systems in chemical and petrochemical plants are used as safety equipment to relieve pressure in plant equipment to avoid explosions and prevent air pollution. These flares are fed at lower flow rates of waste gases than they were designed for. Current EPA regulations limit the exit gas velocity to less than 60 ft/s, but do not consider the lower velocity limit required to ensure complete combustion. Gases exiting a flare tip at low flow rates result in insufficient mixing with ambient air and leading to smoking conditions and excess emission. This reduction causes a decline in combustion efficiency and decreases the flare performance. The main objective of this study is to design and test a new air-assisted flare capable of efficiently operating over a wide range of low flow rates of waste gases. In this work, a flare system consist of a 6-inch diameter have been used to evaluate various flare tip designs capable of operating at low flow rates with high flare performance. A special flare tip has been designed with waste gases to increase the flow rates of these gases. As a result of reducing exit area for waste gases, the flow rate of these gases increases. This work relies on CFD analysis to design the flare tip. Experimental tests have been conducted to validate the CFD model and examine different operating scenarios expected in the field during normal operating.

## 513783 Comparison of Finite Impulse Response Vs. Subspace Vector Identification in Model Identification of a Fractionator

**Abstract:**

Model predictive control (MPC) is oricess model based technology to optimize process performance by manipulating controlled variable setpoints of regulatory control loops. One of the most common package for MPC is dynamic matrix control (DMC) which traditionally uses finite impulse response identification (FIR) method to compute steady-state gains matrix.

While MISO (multiple input single output) structure-based FIR ID is superior for low order parametric models, subspace vector ID implements intensive optimization to become a true MIMO (multiple input multiple output) structure and results in a balance model between steady state gains and short-term dynamics. This work presents a quantitative and qualitative comparison of these two model identification techniques in computing steady state gains matrices by using Aspen DMCplus.

Given fractionator column data, the steady state gain matrices are computed with different amount of data to see the effect of data size on the quality of the resulting model. Further, different levels of noise are introduced to the data to test the effect of noisy data on the quality of steady state gains matrix. The results indicate subspace vector ID has led to better models in term of deviations from the full-blown models based on full amount of data without noises.

## 513818 Controlled Nano-Scale Polymorph Production of Alpha/Beta Glutaric Acid from Aqueous Solution Atomization/Evaporation

**Abstract:**

Glutaric acid is a water soluble and commonly used dicarboxylic acid. It exhibits dimorphism, which means that the molecule can form two different types of crystal structures. The two polymorphs of Glutaric Acid are known as the alpha and beta polymorphs. The alpha polymorph is in the meta-stable phase, meaning that it is less stable and after some time will revert back to the beta polymorph. The alpha polymorph is favored at higher temperatures and is more soluble than the beta phase, and has been produced under some very specific conditions. This is thought to have led to some variability with regard to the data collected in several experiments involving glutaric acid.

In this work, we investigated the formation of alpha glutaric acid polymorph in aerosols generated from an aqueous glutaric acid solution. We have experimented with solution concentration, solution flow rate into the atomizer, and drying rate of the atomized droplets. By manipulating the rate of water evaporation from aqueous solution droplets, we have manged to control which polymorph we produce - alpha, beta, or both at the same time. The formation of both polymorphs from an aqueous solution has never been documented before. Understanding the conditions that lead to the formation of the alpha polymorph will be valuable if this method can be used to produce more pharmaceutical-relevant products, particularly ones that are sensitive to heat or otherwise difficult to control.

# 513851 Particle Scattering Photography Approach for Measurement of Multiphase Reactors: Theoretical Simulation and Experimental Optimization

**Abstract:**

Particle scattering photography approach for measurement of multiphase reactors: theoretical simulation and experimental optimization

Yiting Xiao[1,2], Xiangyang Li[1], Zai-Sha Mao[1], Minghui Xie[3], Chao Yang[1,2*]

[1] *Key Laboratory of Green Process and Engineering, Institute of Process Engineering,*

*Chinese Academy of Sciences, Beijing 100190, China*

[2] *University of Chinese Academy of Sciences, Beijing 100049, China*

[3] *Zhejiang Greatwall Mixers Co., Ltd, Wenzhou, Zhejiang 325019, China*

*Corresponding authors. Tel: +86-10-62554558; E-mail: chaoyang@ipe.ac.cn (C. Yang)

**Abstract**

The local property distributions in multiphase reactors are very important because they can directly reflect the reactor performance and help to profoundly understand hydrodynamics therein. Consequently, they have a major impact on design, scale-up and intensification of multiphase reactors and validation of CFD models. Unfortunately, the existing measuring techniques are ineffective in case of nearly industrial operating conditions (particular physico-chemical environment, opaque walls, high gas holdup or solid concentration, etc.), especially in three-phase systems.

With low-brightness particles as the dispersed phase in multiphase reactors, the photographic measurement is encountered with the great challenge in identifying particle outlines. In this work, a novel approach based on the light scattering principle is proposed to raise the brightness difference and then the photography can be successfully implemented. A single-particle scattering model based on non-coherent scattering hypothesis was developed to evaluate the total scattering light intensity. A series of experiments were conducted and tested the theoretical calculation in turn.

The quantitative relationships between scattering-light-intensity and scattering coefficient/absorption coefficient were established. Further, the partial light fluxes projected onto the focal plane by single particles with different sizes were also calculated. On this basis, four representative oxide powders, i.e., $TiO_2$, $Al_2O_3$, CuO and $SiO_2$ were chosen as scattering particles. Aided by scattering light of trace oxide powders, the grayness differences between two particles-to-be-measured and liquid phases in all images were improved with different degrees and the $TiO_2$ particles performed the best. The grayness differences varied with the $TiO_2$ concentration, which was almost in a normal distribution and the optimum concentration was about 7.3 mg/L. As expected, this particle scattering approach ( <5%) was proved to be more accurate measurement than the reflector-aided one ( >20%).

Some recent industrial application of our measurement, mathematical models and numerical simulations of multiphase reactors/crystallizers have brought about impressive economic benefit by reducing the materials and energy consumption and pollution emissions and saving of equipment investment, which make the industrial processes more sustainable and profitable.

**Keywords**: multiphase reactor; particle scattering photography; measurement; bubble size distribution; particle; concentration

**Acknowledgements**: The National Key Research and Development Program (2016YFB0301701), the Major National Scientific Instrument Development Project (21427814), the National Natural Science Foundation of China (91534117, 21490584), the Key Research Program of Frontier Sciences, CAS (QYZDJ-SSW-JSC030) and the Instrument Developing Project of Chinese Academy of Sciences (YZ201641) are gratefully acknowledged.

***References***

1. G.Q. Wang, X.Y. Li, C. Yang, G. Z. Li, Z.-S. Mao, New vision probe based on telecentric photography and its demonstrative applications in a multiphase stirred reactor, Ind. Eng. Chem. Res. 56 (23) (2017) 6608-6617.

2. A. Amokrane, S. Maaß, F. Lamadie, F. Puel, S. Charton, On droplets size distribution in a pulsed column. pratȼñ: in-situ measurment and corresponding CFD-PBE simulations, Chem. Eng. J. 296 (2016): 366-376.

3. J.A. Boxall, C.A. Koh, E.D. Sloan, A.K. Sum, D.T. Wu, Measurement and calibration of droplet size distributions in water-in-oil emulsions by particle video microscope and a focused beam reflectance method, Ind. Eng. Chem. Res. 49 (3) (2010) 1412-1418.

4. S.F. Yang, X.Y. Li, C. Yang, B. Ma, Z.-S. Mao, Computational Fluid Dynamics Simulation and Experimental Measurement of Gas and Solid Holdup Distributions in a Gas-Liquid-Solid Stirred Reactor, Industrial & Engineering Chemistry Research, 2016, 55(12), 3276-3286.

# SAFETY DEMANDS AND PROPOSAL SOLUTIONS FOR LABORATORIES AND SEARCHING CENTERS IN BRAZILIAN UNIVERSITIES

*Carlos Andre Vaz Junior, Escola de Química / UFRJ, Rio de Janeiro, RJ, Brasil,*
*Alice Souza Gonçalves, UCP, Rio de Janeiro, RJ, Brazil*
*Antonio Carlos de Oliveira Ribeiro, Bayer S/A, Rio de Janeiro, RJ, Brazil*

## Introduction

Laboratories are areas of work and / or teaching used to conduct experiments for purposes of teaching, research or application of scientific processes. This includes chemical, physical, medical, microbiological, genetic engineering laboratories, etc. To perform a safety assessment in laboratories is necessary to associate several safety standards such as Governmental Regulatory Standards Norms (NR) from the Ministry of Labor and Employment (MTE), Fire Protection Code and RAGAGEPs from the Brazilian Association of Technical Standards (ABNT).

Associated with these standards it is common to find professionals involved in laboratory experiments in universities and others publics research centers with poor perception of the risks involved in their activities. Many highly qualified and specialized persons in their field of research, do not have sufficient safety training or education. The absence of a well-developed safety culture in these researchers, coupled with the lack of industry standards and legislation, can lead to deviations, neglect of simple safety rules, and improvisation. These factors have already resulted in serious accidents.

This article shows a proposal for a structured risk analysis methodology to be applied in laboratories. This article aims to cover the main risks involved in the activities carried out in the laboratories with a focus on the methods, technologies and main processes. It is important to note that specific risks, such as those associated with electrical, mechanical or biological hazards, need to be evaluated and if so appropriate improve measures can be defined. The proposal is to present a comprehensive safety analysis for the activities in research and teaching laboratories. This methodology can be applied to chemical, physical, research and pilot testing laboratories.

## Organization measures aplied to laboratories

On the concept of safety education applied to laboratories, based on the Germany BG-RCI safety guide working safely in laboratories (1) , we used four organizational principles: i. hazards identification; ii. risk assessments; iii. minimize risks; iv. emergence response (2) as following defined.

***Hazards identification***

Typical laboratory chemicals are: pure substances, solutions, suspensions or gases that are used as analytical standards, reagents, solvents, etc. Also related are reaction products, by-products and contaminants, including unexpected reaction products.

Materials that do not have hazardous characteristics but have properties that can cause hazards are also considered as hazardous materials. Examples include unstable materials or materials which may result in a risk in case of contact with one another or because of their temperature or thermal capacity, such as hot melt salts or cryogenic liquefied gases. Another significant risk is the choking effect of many gases. Even materials that appear to be harmless can be dangerous, such as cellulose dust, which can explode when mixed with air. Each chemical product need to be evaluated accordingly to the Material Safety Data Sheet (MSDS). For safe experiments, analyzes and procedures it is necessary to have information on the reagents / products used and this information is contained in the Material Safety Data Sheet (MSDS).

Product hazards recognition induces the first stage of safety promotion using inherent safety principles. These principles are based on the concept of: minimizing effects by minimizing the volumes and quantities of dangerous substances handled; substitute hazardous products for less critical products such as the use of toxic or flammable solvents with higher flash point solvents, use of robust equipment; moderate use for less hazardous operating conditions such as refrigeration to result in lower vapor pressure, use of containment; simplify with the use of correct and friendly identifications of products, equipment and installations reducing operational complexity, use of fail-safe installations, that is, in the event of a loss of energy, they fail in a safe position. Further details on inherent safety can be found in the book "Safety of Chemical Processes", by Daniel Crow and Joseph Louvar (3) .

***Risk assessment***

Several techniques for risk analysis can be used in laboratories. To support the decision to use these techniques the American Chemical Society (ACS) has published the guide "Identifying and Evaluating Hazards in Research Laboratories" (2). This guide selects five hazard analysis techniques:

- Chemical Safety Levels - CSL ranking from 1 to 3, with 1 being the least critical. Product characteristics such as levels of flammability of toxicity and other hazards define the safety class and with these classes, the safety rules applicable to the installations and the personnel involved are defined.

- Job Hazards Analysis (JHA) consists of a structured task analysis technique, which associates systematically the relationship between the researcher and the task, taking into account the tools, the necessary equipment and the environment in which task will be executed.

- What – If Analysis. This technique consists of thinking in a structured way about what can be wrong and then judging the probability and consequence of each deviation scenario. It can evaluated by a safety expert or a group. Complementing this technique with the use of a risk matrix for deviations, as used in the HAZOP technique, results in a more robust risk analysis.

- Checklists. A well-structured checklist is an effective tool for risk assessment and implementation of safe practices. We can consider this technique the most recommended for use in laboratory activities. As researchers are familiar with the issues previously presented in the cheklists the time of the risk analysis as well as the time to implement the necessary adaptations is much faster than the other techniques presented. It also allows standardization of risk analysis for a comparison of results between facilities in the organization. With this it is possible to define priorities of actions and investments in the improvement of security performance. In its basic form a security checklist can be the list of major risks related to materials, activities and equipment. This technique can be complemented by the others afforded mentioned techniques.

- Standard Operations Procedures- SOP. Each step of the experiment in the laboratory should be analyzed separately to identify points of failure and then determine possible combinations of these faults and their impacts. The advantage of this technique is that it can be done by any lab worker. A list of topics is performed to see how each safety topic is covered in the SOP.

***Minimize Risks***

Laboratory activities, in particular involving hazardous materials, require structural, technical, organizational and personal protective measures.

With regard to the ranking of protection measures, technical measures take precedence over organizational and personal measures. We can have a road map to check the risks involving:

- Personal protection measures;
- Safety in constructions, design and equipaments;
- Ergonomical aspects
- Electrical safety
- Pipelines and conections
- Fire protection
- Static electricity hazards
- Explosion protection hazards
- Ventilation
- Storage of dangerous products
- Equipment operation and integrity with considerations about : flexible hoses and connections, gas burners, drying systems, hot surfaces, pressure vessels, distillatory, refrigerators, Dewar vessels, heating baths, thermal fluids, heating ovens, muffles, centrifuges, fume hoods design and operations,
- Organizational measures

***Emergency response***

Being prepared for emergency actions in case of accidents should be mandatory for anyone handling hazardous products in laboratories. Understanding what to do in an emergency is essential. How to use combat and emergency equipment, where these equipment are located. Know what to do in case of a chemical leak

Risk assessment should also include emergencies and failures. The ability to control events and the impact of individual workstations on one another are particularly important in large laboratories. For example, additional containment systems may be required in hoods if there is a fire.

Simulate the use of a fire extinguisher, such as wiping up spilled liquid, using eye wash and emergency shower, etc. Appropriate safety drills should be performed at regular intervals to ensure that the system works effectively in an emergency.

Special attention should be given to external resources for emergency, such as the fire department. It should be possible to take the necessary safety measures in a timely manner. Clear instructions and guidance for first aid and an agreed place for the victim dispatch in case of accidents should be available.

## Conclusions

Process safety has developed strongly in the last decades, safety in teaching and research laboratories has generally been seen as less important. There is often a view that laboratory accidents will always be smaller and less significant. Unfortunately, this vision does not always materialize. The final report on the explosion at Texas Tech University, prepared by the Chemical Safety Board (CSB) (4) clearly demonstrates that risks exist and should be considered.

This paper focused on the construction of basic pillars to ensure safe operation in a laboratory environment. Hazard recognition, risk and hazard assessment, risk minimization and emergency preparedness were addressed.

It is under preparation a CCPS guideline for Laboratory Safety ("Guidelines for Process Safety in Chemical Laboratories and Pilot Plants") (5). By this time we assume the use of German BG-RCI safety guide "Working safely in laboratories" (1), is the best available reference to handle with risks involving laboratories installations.

## References


1. Working Safely in Laboratories, "213-851", BG RCI- DGVU, (2014).

2. Identifying and Evaluating Hazards in Research Laboratories Guideline, American Chemical Society, (2015).
Available in:
https://www.acs.org/content/dam/acsorg/about/governance/committees/chemicalsafety/publications/identifying-and-evaluating-hazards-in-research-laboratories.pdf

3. Safety of Chemical Processes, Daniel Crow and Joseph Louvar. 3th ed. (2015).

4. Texas Tech University Chemistry Lab Explosion, CSB – Chemical Safety Board, (2011).
Available in http://www.csb.gov/texas-tech-university-chemistry-lab-explosion/

5. "Guidelines for Process Safety in Chemical Laboratories and Pilot Plants", CCPS, Wiley
Information available in
https://www.wiley.com/en-us/Guidelines+for+Process+Safety+in+Chemical+Laboratories+and+Pilot+Plants-p-9781119010135

# 512892 Experimental Study of Gas Leakage on Pressure Fluctuation in a Cyclone Separator

**Abstract:**

In order to get further insight into characteristics of pressure affected by the gas leakage from dipleg in a cyclone, pressure drop and pressure fluctuation were measured by U-tube and multi-input dynamic pressure transducers respectively. The results show that in gas - particle two-phase flow the pressure drop decreases first, after increases, at last decreases sharply with the increase of gas leakage. However, the pressure drop decreases linearly with the increase of gas leakage in gas phase. Pressure signals at different tappings were measured by dynamic pressure transducers. By adopting, standard deviation, mean value, and power spectral density methods, the characteristics of the pressure signals were analyzed and discussed. Experimental results show that tapping 2 of the outlet and tapping 1 of the inlet were affected little by the gas leakage, tapping 6 of dipelg leakage average pressure was affected greatest, and intersection points which were the maximum of pressure fluctuation and main energy from low frequency were the average pressure curves tapping 6 and different axial tappings intersect at. The pressure drop of the cyclone separator is reduced, the transient pressure fluctuation is enhanced, and the instability of the flow field is increased by the gas leakage of dipleg. The experimental results can provide deepen insight into the variation of pressure field in a cyclone separator with gas leakage.

## 513642 Senior Capstone Design Project for the University of South Florida

**Abstract:**

The senior capstone class at the University of South Florida has 40 design project groups addressing five themes; fuel cells, hydrocarbon processing, waste water treatment, biomass conversion and micro encapsulation and timed release products. The best senior capstone project that involves simultaneous design of process and products for a fuel cell application will be presented.

# 514986 Thermal Evaluation and Assessment of Process Burners using CFD

**Abstract:**

In the past few years, CFD numerical methods have improved and became a favorable tool for many thermal types of equipment. For example, the tool was used here in this work, for the development and optimization of process burners used in petrochemicals industries such as (ethylene steam crackers, reformers, and EDC furnaces). The focus of this work was to predict the burner flame structure and distributions of velocities, temperature, and emissions inside the furnace volume. Detailed chemistry mechanisms, radiation, and turbulence were accounted in the model that resulted in very reasonable predictions of the flame structures at low computational cost. The results showed acceptable agreements of flue gas concentrations compared with burner plant and design data. However, for further accuracy in the prediction of emissions like NOx and CO, advanced turbulence model need to be incorporated to resolve most of the kinetic energy that can directly affect scalar mixing of spices, this will come at higher computational cost like in LES or DES turbulence models.

## 521161 Industrial-Academic Initiatives, Aligning Needs and Building Community

**Abstract:**

AIChE industrial-academic initiatives are in need of actionable outcomes that support collaborative success and developing professional researcher goals. Given that the Spring AIChE meeting and Global Conference of Process Safety is the setting for industrial practice, it is the natural setting for academics to participate in towards building relationships inline with such initiatives. This poster shares research experiences and curriculum development aligned with calls from students, industry needs, and global policy over the last decade. Focus of the poster is placed on energy, natural gas technology-education, international development, internet of things, informatics, device development (with a focus on sensors), and green chemistry-engineering. The presenter encourages potential industrial partners to stop by in the hopes to establish a dialogue of how to build relationships that support industrial needs and researcher development. Further, AIChE leaders interested in championing and spotlighting this aspect of the Chemical Engineering community are also welcomed to stop by.

# 512075 A New Process for Poultry Manure Revalorization: Fresh Water and Thermal Energy Savings with Effluent Recycle, Biogas Production, Thermal Hydrolysis, Anaerobic Digestion and Snad Process

**Abstract:**

Valorization of poultry manure through the anaerobic digestion (AD) generates an effluent with high ammonia concentration that is inhibitory of the AD process. This effluent must be treated for its subsequent elimination. Biogas production favors AD, though in different countries AD was implemented on a relatively large scale only when green gas subsidies were guaranteed. This is due to the high cost of fresh water and heating of bioreactor units. In addition, due to the high viscosity of manure, a pretreatment to facilitate its metabolization to methane is required. Thermal hydrolysis can be used as pretreatment of AD to accelerate the metabolization, to increase the volume of biogas produced and to decrease the amount of solids to be disposed. In addition, the process of nitrogen abatement comprising the simultaneous partial nitrification, Anammox and denitrification is capable of removing nitrogen species. Then, the effluent can be used for dilution of the poultry manure producing cost savings in fresh water and heating. However, the coupling of these units and the evaluation of the reuse of the SNAD effluent as a solvent has not been evaluated on an experimental scale.

The aim of this work was to study the effect of recycling the final effluent resulting from the stepwise process of Thermal Hydrolysis (TH) of poultry manure combined with AD and simultaneous partial nitrification, Anammox and denitrification (SNAD).

Thus, the proposed system was a coupled process of TH+DA+SNAD. First, poultry manure was subjected to a TH treatment, which solubilizes the particulate organic matter and removes the calcareous material (calcium carbonate) from the slurry. The poultry manure was heated at 180 °C and 20 psi during 3 hours; a filtration step was added, which allows the removal of larger solids. The resulting viscous slurry was first diluted with tap water, then with the final effluent of the SNAD process and anaerobically digested in a 3.5 L granular reactor. Finally, a SNAD reactor (4.3 L) previously set and operated for several months, was stepwise fed with the effluent from the DA diluted with tap water until an inlet concentration of 400 mg total ammonia nitrogen (TAN)/L was reached. As a last step, the effluent from the SNAD process was recycled to the anaerobic reactor. The effect of the recycle was evaluated through the analysis of the coupled process efficiencies and the biomass characterization (relative abundance of populations, specific activities, sedimentation rate, sludge volumetric index and granules size).

Efficiencies up to 81% of COD removal were obtained in the anaerobic digestion process using thermal pretreatment in the poultry manure. In the SNAD process fed with the effluent from the anaerobic digester, a total ammonia nitrogen and total inorganic nitrogen removal of 100% and 72%, respectively, were obtained. Considering a stoichiometric efficiency of 89% in the removal of nitrogen by the partial nitrification-Anammox process, a good performance of the reactor was reached. When recycling the effluent from the SNAD process to the DA (recycle ratio of 0.32), the efficiency of the anaerobic digester was 65% and the methane percentage in the biogas was 73%. The efficiency decrease was explained by the decrease of the COD concentration in the influent, which is probably due to the presence of nitrate in the recirculation liquid, which in turn can be reduced to nitrogen gas by denitrification when organic matter is present. Despite the removal decrease, the methane content in the biogas of the DA increased from 62% to 73% in the recycle condition. Then, the system was stable when recycling in terms of the COD concentration of the effluent and in the biogas production. On the other hand, in the SNAD system the specific Anammox and nitrifying activities were 0.210 $gN\text{-}N_2$ /gSSV • d and 0.344 $gN\text{-}NH_4^+$ /gSSV•d using recycle, and 0.501 $gN\text{-}N_2$/gSSV•d and 0.057 $gN\text{-}NH_4^+$/gSSV•d without recycling. The sludge volumetric index was 26 mL/gSSV; the average equivalent diameter of the granules with and without recycle was 1.285 and 1.515 mm, respectively. The relative bacteria abundance did not show a significant difference in relation to the recycle and non recycle condition, showing mostly groups of aerobic ammonium oxidizing bacteria (aeAOB) and anaerobic ammonium oxidizing bacteria (anAOB) with values above 79%.

In conclusion, The TH-DA-SNAD coupled process with recycling allows to obtain a high removal of nitrogen, without substantially changing the relative abundance of the different bacterial communities; increases specific nitrifying activity and Anammox; decreases the average equivalent diameter of the granules and allows to obtain a good sludge volumetric index. Further challenges in order to optimize the proposed process are: to increase ammonia concentration in the influent of the SNAD system or to replace the TH with a lower operational cost unit.

Acknowledgements: This study was made possible by FONDECYT (Chile) [Grant Number 1140491] and INNOVA (Chile) [Grant Number 15VEIID-45613].

## 512942 Future of Mobility: A Driver for Change in Energy and Chemicals

**Abstract:**

Future mobility will be impacted by a drive to low- or zero-emission vehicles to mitigate smog, and accommodate urban migration. Climate change impacts from $CO_2$ emissions are a global concern. Autonomous vehicles, if adopted, will change patterns in vehicle use and ownership. Optimal mobility options light duty vs. heavy duty transit, and for aviation may require different solutions. This poster will seek to engage discussion on preferred options, and discuss impact on the chemical engineering profession for providing energy, fuels, and chemicals.

# 512134 Safety Improvement and Thermal Management of Cylindrical Lithium-Ion Batteries Using Phase Change Materials (PCMs)

**Abstract:**

**Safety Improvement and Thermal Management of Cylindrical Lithium-ion Batteries Using Phase Change Materials (PCMs)**

1. M. Sadrameli*, Y. Azizi,

Process Engineering Department, Faculty of Chemical Enginering,

Tarbiat Modares University (TMU), Tehran, Iran.

**Abstract**

Lithium-ion batteries (LIBs) have shown to be one of the most reliable energy storage systems for the electric cars and motorcycles in the recent years. Ambient temperature has a significant impact on the performance, life time, safety and cost of such batteries. Increasing the surface temperature during the charging and discharging degrades the lithium batteries more quickly while working at low temperature environment results reducing the power and energy capability of the system. A thermal management system has been designed and setup in laboratory scale for controlling the temperature at optimum conditions using PEG-1000 with melting point in the range of 33-40 $^{o}$C as a phase change material. Aluminum plates have been installed in the PCM to increase the thermal conductivity and increasing the heat transfer rate. Experimental tests have been run at different discharge rates and ambient temperatures to investigate the effects of temperature on the efficiency of the batteries. The comparison has been made between the system of 6 batteries with and without PCM and the results show that PCM with aluminum plates decrease the surface temperature of the batteries by 19%, 21%, and 26% for the rates of 1 C, 2C, and 3 C respectively that would results better performance and longer life time of the batteries.

Keywords: Lithium-ion batteries; thermal management; temperature control, phase change materials

- *Corresponding author, Email:sadrameli@modares.ac.ir*

# 512107 Dissolution Mechanism of Silica and Alumina Minerals By the Chemical Leaching of High Ash Indian Coal


**Abstract:**

Coal is the most abundant fossil fuel and cheapest energy source in the world. Coal plays a vital role in the global energy demand, especially through power generation. The hard coal reserves are gradually depleting, and the mining operation at deeper coal seam with greater difficulties as well as the cost of exploration is so high, which has a significant issue for plant economics. So to minimize the problems instead the low grade coal can be used as an alternative energy source. Low-grade coals could be important to both the energy demand and economic development for the industries. Chemical cleaning techniques is an efficient method for to remove the organic and inorganic minerals without affecting the carbon content of coal. A method was devised for the removal of deleterious minerals and trace elements from the low-grade coals. An attempt has been made in this investigation to study the effect of leaching concentration and kinetics of dissolution major minerals constituent (especially silica and alumina) from low-grade coal was studied. The low-grade coal was selected from Mahanadi Coalfield of Odisha, India. The coal sample (250 μm particle size) was treated with 10 to 40 by wt% of caustic concentration at 60 °C to 120 °C for 2 h. The investigation results showed that the degree of demineralization was improved with increase concentration; which reduces the ash content above 50% by the caustic leaching effect. Apart the leaching study, the chemistry of demineralization was evaluated from the X- ray diffraction (XRD) analysis and X-ray fluorescence (XRF) spectrometer. Throughout the research study challenges the chemical cleaning of low grade coal has been efficient techniques for reducing the minerals to a minimum level that can be upgraded to high rank coal.

# 513343 Combination of Sea and ED Methods for Preparation of Active Bimetallic Catalysts for Direct Methanol Fuel Cells

**Abstract:**

Direct methanol fuel cells (DMFCs) have received widespread attention because of their potential applications in transportation and portable electronic devices. Platinum (Pt) catalysts are effectively able to catalyze the oxidation of methanol to carbon dioxide in this system. The main drawback for Pt-based catalysts is that the CO intermediate chemisorbs strongly on Pt surface; even traces of CO will lead to a severe poisoning of Pt catalysts, which prevents further catalyst turnovers. From previous studies, bimetallic catalysts composed of platinum (Pt) and a second metal such as Ru or Ni are more active for the oxidation of methanol compared to catalysts composed only of Pt. In addition, the stability and activity of catalysts depend directly on the architecture, which is greatly affected by both the composition and preparation methods [1]. In this study, nanometer sized Pt-Ru and Pt-Ni particles on a carbon support are prepared by combining the methods of strong electrostatic adsorption (SEA) and electroless deposition (ED).

The method of SEA has been used to form supported Pt nanoparticles with high dispersion and narrow size distributions. Catalysts prepared by SEA are then used as seed nuclei for addition of a secondary metal (Ru or Ni) using Electroless Deposition (ED). Thus, two series of carbon- supported bimetallic catalysts were prepared containing controlled and variable coverages of Ru or Ni on Pt surfaces, which were then characterized and evaluated for methanol oxidation reactions. These sets of catalysts show higher activity towards methanol oxidation in comparison to the analogous, commercial catalysts.

**References**

1. Ehteshami, S.M.M. and S.H. Chan, A review of electrocatalysts with enhanced CO tolerance and stability for polymer electrolyte membarane fuel cells. Electrochimica Acta, 2013. 93: p. 334-345.

## Effects of Fossil Fuel $CO_2$ Emissions on the pH of the Oceans

*Dr. Frank J. Elvin, Texas Specialty Catalysts, Bellaire, TX*

### Abstract

Surface seawater samples from the Pacific, Atlantic, Gulf of Mexico, North Sea, Baltic Sea, Black Sea and the St. Lawrence Seaway were tested in-situ and under laboratory conditions for pH, $CO_2$ vapor pressure, and rates of $CO_2$ absorption. Laboratory tests were conducted on each sample under atmospheric $CO_2$ partial pressures of between 0ppm and 4,000ppm. The pH of every sample = A – B.$\log_e$($CO_2$ vapor pressure), where A represents the pH when the sample $CO_2$ vapor pressure is 1ppm, and B is the slope of the logarithmic function. A is approximately 10.6 and B is approximately 0.4. The $R^2$ correlation for each sample was about 0.98. The in-situ tests confirmed National Oceanic and Atmospheric Administration (NOAA) data showing that $CO_2$ vapor pressure averaged 7% lower than the atmospheric $CO_2$ partial pressure. This difference is the driving force for $CO_2$ absorption by the oceans. The in-situ seawater pH ranged from 7.9 to 8.5 corresponding to $CO_2$ vapor pressures of 410ppm to 250ppm. The average pH was 8.08 with a $CO_2$ vapor pressure of 378ppm. This study shows that the average surface seawater pH would decrease to 7.8 if the atmospheric $CO_2$ increased to 800ppm and to 7.2 if $CO_2$ increased to 4,000ppm. The increase in seawater acidity may be explained solely by the increases in fossil fuel $CO_2$ emissions. The average ocean pH could be maintained at 8.08 by reducing fossil fuel $CO_2$ emissions to 15Gt/year from approximately 35Gt/year. The longer this reduction in fossil fuel emissions is delayed, the lower the ocean's pH will drop. The differential $CO_2$ concentration at the ocean surface results from the absorption and reaction of the dissolved $CO_2$ by plants, and the change in chemical equilibrium. This paper examines the causes of the observed variation in differential $CO_2$ concentration at the ocean's surface.

### Introduction

In 2011 the following equation was developed to represent the annual change in atmospheric $CO_2$ concentration. $\delta CO_2 = \delta Ef + 5.0 - 0.017CO_2$ where $\delta CO_2$ is the increase in atmospheric $CO_2$ concentration in ppm in a calendar year, $\delta Ef$ is the fossil fuel $CO_2$ emissions in ppm in that year, and $CO_2$ is the average atmospheric carbon dioxide concentration in January of that year. This equation is based on mass transfer theory and reaction kinetics described by Levenspiel (1) and Lewis and Whitman (2). The values of pre-1958 atmospheric $CO_2$ concentrations were from ice-core samples (3), and the values of post-1958 atmospheric $CO_2$ concentrations were measured at Mauna Loa (4). Annual fossil fuel emissions $\delta Ef$ were from Energy Information Administration (EIA) (5), Carbon Dioxide Information Analysis Center (CDIAC) (6) and International Energy Agency (IEA) (7). The above equation was accurate through 2017 and it was determined that the two equation constants had not changed between 1750 and 2017. The prediction of atmospheric $CO_2$ concentration between 1750 and 2017 is shown in figure 1. The accuracy of these predictions indicated that fossil fuel emissions were the only significant contributor to the rising atmospheric $CO_2$ concentrations, and the net oceanic absorption was proportional to the atmospheric $CO_2$ concentration. These conclusions were supported by measurements of seawater $CO_2$ vapor pressures and $CO_2$ absorption rates (8).

This paper establishes the relationship between ocean surface water pH and the water $CO_2$ vapor pressure. It also establishes the relationship between seawater pH and the atmospheric $CO_2$ concentration in the air above the seawater. A high accuracy portable $CO_2$ meter from CO2meter.com measured atmospheric $CO_2$ concentrations above the seawater.

Average ocean pHs are predicted through 2100 for different rates of fossil fuel $CO_2$ emissions.

Supplementary Excel and Word files (not attached to this paper) contain all data, calculations and predictions. They also include statistical and sensitivity analyses and detailed methodology.

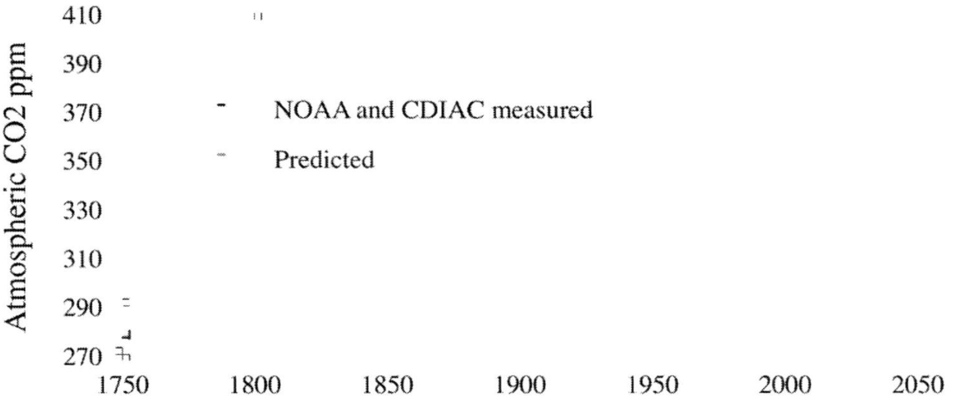


Figure 1 Predicted Atmospheric CO2 1750 to 2017

**Absorption and desorption of $CO_2$ by seawater**

The oceans are not in equilibrium with the atmosphere. It is this lack of equilibrium that facilitates $CO_2$ absorption. Without photosynthesis the oceans' surface waters would usually be in equilibrium and the $CO_2$ vapor pressure would equal the atmospheric $CO_2$ partial pressure. However, changes in temperature can result in the water being out of equilibrium, even without photosynthesis, but if this happens, $CO_2$ absorption or desorption will bring them back. Any diurnal or seasonal temperature effects are assumed to be insignificant and self-canceling over the calendar year time frame used for calculating average absorption rates and rate constants.

Photosynthesis removes $CO_2$ from the surface water, whereas absorption from the atmosphere, biomass respiration, decay and decomposition add $CO_2$. These mechanisms are functions of $CO_2$ concentration, temperature, sunlight availability and intensity, and the density of surface plants, plankton and nutrients. Some of the carbohydrates produced by photosynthesis are ultimately converted to calcium carbonate in shells, fish bones, and coral reefs. The global average $CO_2$ vapor pressure *($VPCO_2$)* was approximately 93% of the atmospheric $CO_2$ partial pressure *($PCO_2$)* between 1990 and 2012 NOAA (9). The atmospheric $CO_2$ partial pressure "$PCO_2$" used in this section of the paper is similar to the atmospheric $CO_2$ concentration "$CO_2$" used in the introduction. Both use ppm as their unit of measurement. The average measured difference between $PCO_2$ and $VPCO_2$ was 7% for the ten sample locations listed in figure 3. Samples were tested in 2017 and 2018. The absorption/desorption rate and $VPCO_2$ was determined experimentally using a sealed vessel containing the water sample with a pH meter, and a gas phase $CO_2$ meter. The vessel atmospheric volume was $3.9\times10^{-3}$ $m^3$, the water volume was $1.0\times10^{-3}$ $m^3$ and water surface area was $3.6\times10^{-2}$ $m^2$. All absorption and desorption tests were done at 25°C. The initial $PCO_2$ in the vessel was between 50ppm and 4,000ppm depending on the test. As time passed during each test, $PCO_2$ increased if the initial $PCO_2$ was lower than $VPCO_2$ and decreased if the initial $PCO_2$ was above $VPCO_2$. The $CO_2$ meter in the sealed vessel measured $PCO_2$. The absorption/desorption rate was $k_a.a(PCO_2 - VPCO_2)$ for all samples. The absorption rate constant $k_a$ was calculated from measured values of r, $PCO_2$, $VPCO_2$, $PCO_2$ and a (absorption area). The rate constant was similar for samples from all locations. An absorption and desorption plot is shown in figure 2 for a seawater sample from the Gulf of Mexico at the Galveston San Luis Hotel breakwater. The $k_a$.a was –0.14ppm/hour/ppm. The $CO_2$ in the sealed container could be calculated from $PCO_{2\,at\,time\,t} = e^{-}$

$^{0.14t}(PCO_{2c\ initial} - PCO_{2equilibrium}) + PCO_{2equilibrium}$ where "t" is time in hours of absorption or desorption.

The equivalent global oceanic rate constant $k_a$ was -4.57E-08mt/year/$m^2$/ppm and $k_a$.a was – 0.0047ppm/year/ppm. The global rate constant was calculated from the measured rate constant assuming the ocean area was $3.67 \times 10^{14}$ $m^2$ and the volume of the Earth's atmosphere was $3.97 \times 10^{18}$ $m^3$ (mass of $5.14 \times 10^{18}$ kg). Different ocean surface areas were used to estimate $k_a$.a. The surface area most representative assumes that the effective global surface area is 2.24 times the actual flat or calm water area. This is equivalent to the world's oceans being covered with 2 foot high waves, each with a wavelength of 8 feet. In addition, each 2 foot wave is assumed to be covered with 2 inch high ripples each with a wavelength of 8 inches. This assumption is based on observations made while taking the seawater samples.

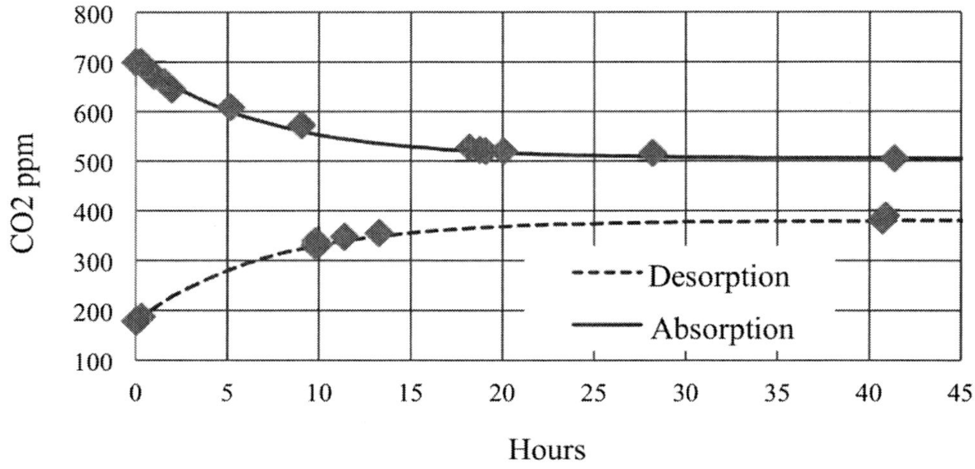


Figure 2 Absorption and desorption of CO2 in sealed container Galveston San Luis

**Ocean pH as a function of seawater $CO_2$ vapor pressures at 25°C**

During seawater testing to measure $CO_2$ vapor pressure ($VPCO_2$) it was observed that $VPCO_2$ could be derived directly from seawater pH, saving many hours of testing per sample, and minimizing post sampling contamination. The pH vs. $VPCO_2$ is shown in figures 3, 4 and 5 for normal salinity (34-36g/kg) and low salinity (15-20g/kg). The Baltic and Black seas were the only low salinity samples. As pH is the log function of $H^+$ ion concentration, pH of seawater was a logarithmic function of $VPCO_2$. The average pH of normal salinity samples was $-0.42Ln(VPCO_2) - 10.55$ whereas the pH of the low salinity samples was $-0.39Ln(VPCO2) - 10.57$. The pHs of some recent samples from Miami Beach, Bahamas, Caribbean, Hawaii, and Antarctica all fit on the normal salinity curves. It is assumed that the low salinity samples have higher pHs because the ratio of $Ca^{++}/Cl^-$ and $Ca^{++}/Na^+$ is 20% higher for the low salinity samples (11). $PCO_2 - VPCO_2$, or $\Delta CO_2$ may be calculated from ΔpH using figure 6 for Galveston seawater. $\Delta pH = pH_e - pH_i$, where $pH_i$ is the surface water pH measured in-situ and $pH_e$ is the sample pH after it has equilibrated in the atmosphere at the sample site. If a $CO_2$ meter is not available, $PCO_2$ and the equilibrium $VPCO_2$ can be estimated from pH using figure 6. This is an exponential curve $\Delta CO_2 = 397 - e^{((\Delta pH - 2.5)/-0.42)}$ .It may be approximated as linear over the range of pH measured. The linear equation is $\Delta CO_2 = 766\ \Delta pH + 3.8$

Figures 3,4, 5 and 6 were derived at 25°C. Preliminary tests show that figures 3,4 and 5 accurately predict pH from vapor pressure between 20°C and 30°C. Figure 6 predicts pH accurately for temperatures as low as 4°C.

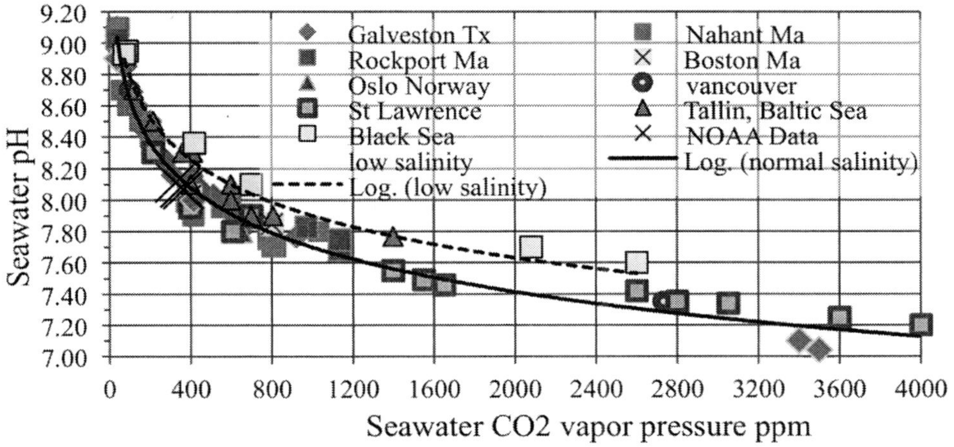


Figure 3 pH vs. Seawater CO2 vapor pressure

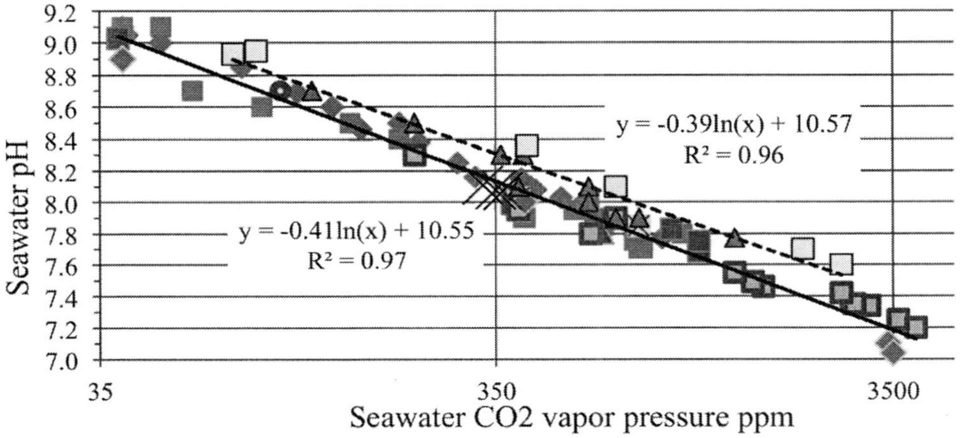


Figure 4 pH vs. Seawater CO2 vapor pressure

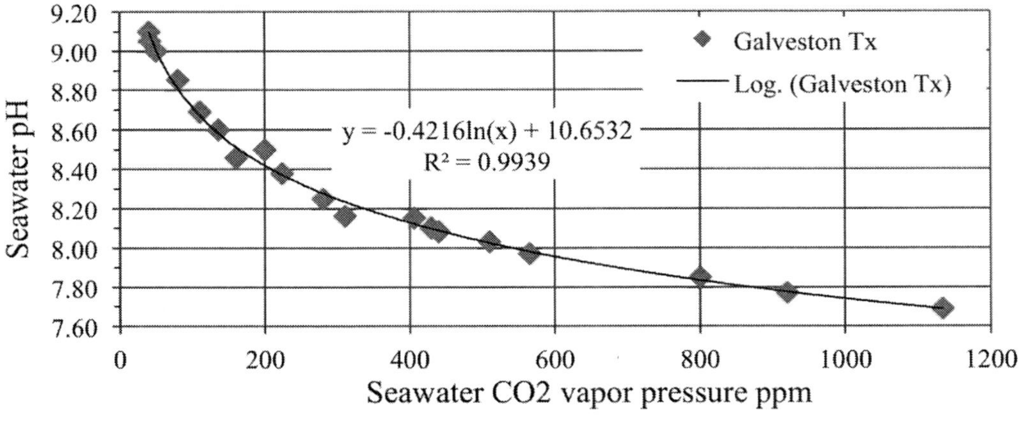


Figure 5 pH vs. CO2 vapor pressure

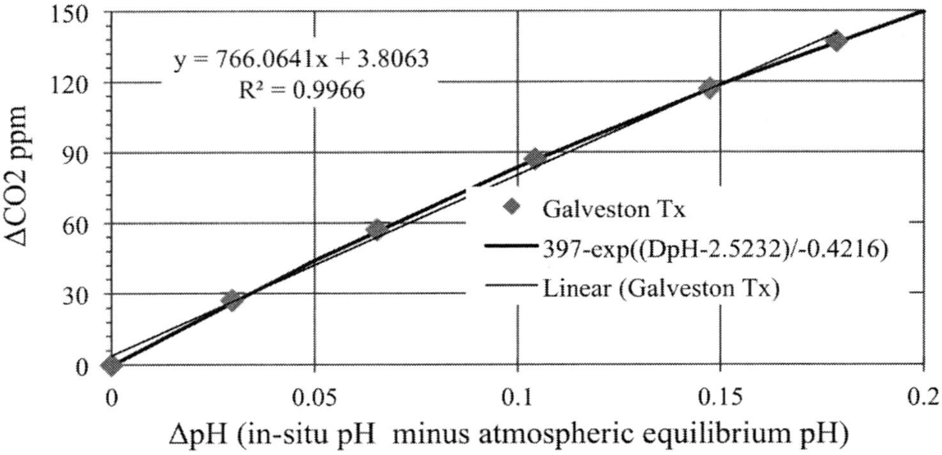


Figure 6 ΔCO2 (PCO2 - VPCO2) ppm vs. ΔpH

**Atmospheric $CO_2$ concentrations and ocean pHs predicted through 2100**

Figures 3, 4 and 5 show that seawater pH is a function of seawater $CO_2$ vapor pressure. The measurement of atmospheric $CO_2$ and $CO_2$ vapor pressure at locations listed in figure 3, show that the average difference between the measured $CO_2$ vapor pressure and the atmospheric partial pressure was 7%. This agreed with the NOAA published results (9). As atmospheric $CO_2$ partial pressure increases, the difference between atmospheric $CO_2$ and $PVCO_2$ increases because photosynthesis increases (10). The seawater pH for a given atmospheric $CO_2$ partial pressure or concentration is approximately the same as the pH of the seawater with a vapor pressure of 93% of the atmospheric partial pressure. Before predicting the oceans' pHs it was necessary to predict the atmospheric $CO_2$. From 2013 to 2015, global fossil fuel $CO_2$ emissions were nearly constant at 33Gt/year although they increased to 35Gt/year in 2016. Figure 7 shows predicted atmospheric $CO_2$ for several scenarios. If fossil fuel emissions remain at 33Gt/year from 2018 through 2100, and there are no changes in other factors, atmospheric $CO_2$ will be 501ppm in 2100. Atmospheric $CO_2$ will ultimately reach equilibrium at 529ppm. However, if annual fossil fuel emissions were to increase at 1.8%/year, which was the average annual increase in fossil fuel emissions between 1980 and 2010, atmospheric $CO_2$ would be 838ppm by 2100 and continue to increase. A 54% reduction in fossil fuel emissions from 33Gt/year to 15.16Gt/year in 2018, and maintaining 15.16Gt/year indefinitely, atmospheric $CO_2$ will equilibrate at 400ppm. If fossil fuel $CO_2$ emissions were eliminated in 2018, $CO_2$ would be 315ppm in 2100 and would equilibrate at 290ppm.

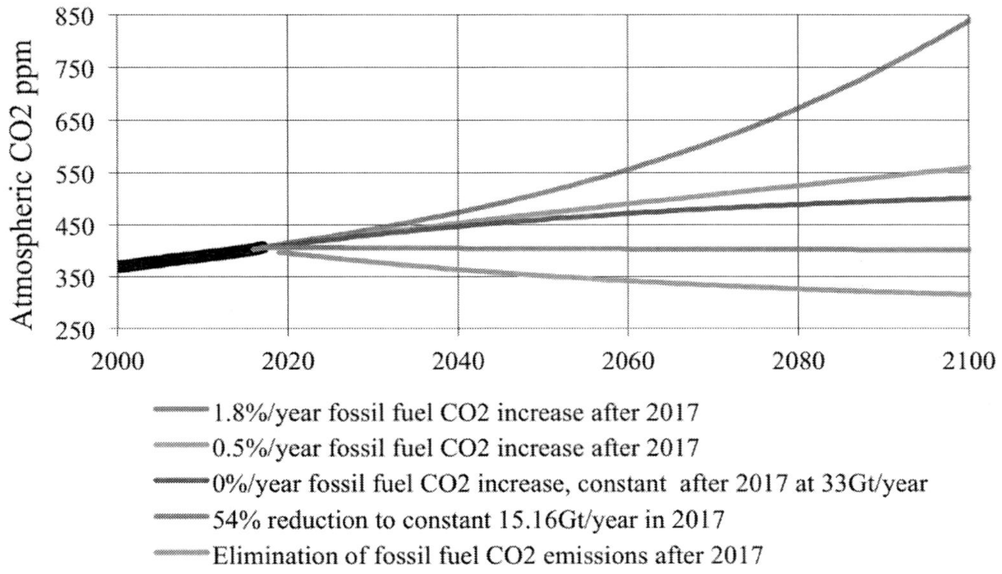


Figure 7 Atmospheric CO2 concentrations: 2000 to 2100

The seawater $CO_2$ vapor pressure $VPCO_2 = 0.93PCO_2$. Therefore pH equals $-0.42Ln((PCO_2)0.93) - 10.55$. Figure 8 shows the pH predictions from through 2100.

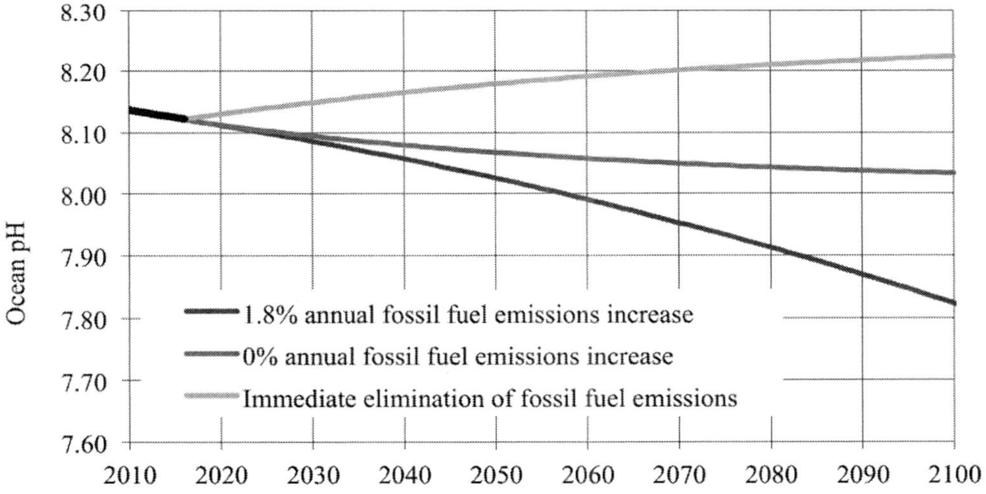


Figure 8 Average pH of 35% Salinity Oceans: 2000 to 2100

## Conclusions

The net rate of $CO_2$ removal from the atmosphere between 1750 and 2016 was proportional to the atmospheric $CO_2$ concentration and increased as atmospheric $CO_2$ increased. There were no significant changes in the average atmospheric $CO_2$ absorption rate constant, or total absorption surface area between 1750 and 2016. There was a 33Gt/year increase in fossil fuel $CO_2$ emissions, which caused 89% of the 123ppm increase in atmospheric $CO_2$, and a 1.95Gt/year increase in non-fossil fuel $CO_2$ emissions, which caused the other 11% of the increase. The average oceans' pH decreased form 8.28 to

8.12 between 1750 and 2017. Without these increases in $CO_2$ emissions, atmospheric $CO_2$ would have remained at 277ppm and the pH would have remained at 8.28. If $CO_2$ emissions were reduced to their 1750 level, atmospheric $CO_2$ would return to 277ppm and the oceans' pH would return to 8.28. This would take approximately 200 years because, as atmospheric $CO_2$ decreases, the rate of $CO_2$ removal also decreases. If $CO_2$ emissions were reduced by 17.84Gt/year, atmospheric $CO_2$ concentration would equilibrate at the 2015 level of 400ppm. This may be achieved by reducing fossil fuel $CO_2$ emissions from 33Gt/year to 15.16Gt/year. If fossil fuel emissions revert to their pre 2011 1.8% annual increase, the oceans' average pH would decrease to 7.8. This would have a serious effect on the oceans' ecosystem and encourage species to evolve that can survive in the lower pH environment.

If future $CO_2$ mass transfer dynamics and kinetics are similar to those from 1750 to 2017, and non-fossil fuel emissions continue to change slowly, fossil fuel $CO_2$ emissions will be the main variable affecting atmospheric $CO_2$ levels and oceans' pH. The reduction of fossil fuel emissions is the only effective method of increasing and stabilizing the oceans' pH.

**References**


1. Octave Levenspiel, Chemical Reaction Engineering, Wiley & Sons 1964 p386
2. W. K. Lewis, W. G. Whitman Ind. Eng. Chem., 16, No. 12 1924
3. Carbon Dioxide Information Analysis Center (CDIAC) Historical $CO_2$ record from the Law Dome DE08, DE08-2, and DSS ice cores June 1998 D.M. Etheridge
4. National Oceanic and Atmospheric Administration (NOAA) Earth System Research Laboratory ESRL atmospheric $CO_2$ data published July 16 2012. Data added for 2012-2017
5. Energy Information Administration (EIA) International Reports 2006, 2009, 2011). International Energy outlook 201. 3 July 35 2013. Report Number: DOE/EIA-0484(2013)
6. Carbon Dioxide Information analysis Center (CDIAC): Global $CO_2$ Emissions from Fossil-Fuel Burning, Cement Manufacture, and Gas Flaring: 1751-2010 July 30, 2013 Tom Boden, Carbon Dioxide Information Analysis Center Oak Ridge National Laboratory
7. International Energy Agency (IEA) May 24 2012 Newsroom
8. The Effects of Fossil Fuel Emissions and Seawater Absorption on Atmospheric Carbon Dioxide concentrations. Presented at the AICHE 2018 Spring Conference
9. NOAA's PMEL Carbon Program (www.pmel.noaa.gov/CO2)
10. Photosynthesis and plant growth at elevated levels of $CO_2$ Amane Makino and Tadahoko Mae. Plant Cell Physiol. 4-(10)):999-1006 (1999)
11. Calcium Distribution in the world ocean waters, B. S. Krumgalz, OCEANOLOGICA ACTA 1982 Vol. 5 N° 1

# 513676 Simulation of Atmospheric Cr Chemistry during Air Sampling (pH 9)

**Abstract:**

Cr(III) is crucial microelement for living organisms and exhibit important role in control of lipids, glucose and proteins metabolism. In comparison, Cr(VI) is highly toxic and US EPA considered as â€˜Group A carcinogensâ€™ that leads to asthma, lung cancer, bronchitis, pneumonitis, nasal damage. Thus, it is important to be able to distinguish between these two oxidation states in the atmosphere[1]. Cr is produced mainly from anthropogenic sources including Cr alloy processing, ferrochrome industry, welding, etc. When particles absorb water (deliquescence), a liquid layer forms around the particles where soluble species including soluble Cr dissolves in the liquid film. Consequently, the inter-conversion between Cr(III) and Cr(VI) can take place with atmospheric reactants and oxidants under atmospheric PM (Seigneur Christian and Elpida Constatinou, 1995). It was found that atmospheric Cr(III) is oxidized to Cr(VI) in the presence of Mn(III) at atmospheric conditions, while Cr(VI) is reduced to Cr(III) in the presence of atmospheric metal ions such as Fe(II), $VO^{2+}$ and As(III) (Amouei Torkmahalleh et al., 2013). This inter-conversion between Cr(III) and Cr(VI) results in some biases in sampling and analytical measuring at pH=9 atmospheric Cr concentrations (Lihui Huang, 2013). The objective of this study was to develop a conceptual model to better understand the inter-conversion for soluble and insoluble forms of Cr compounds in the presence of PM matrix and dissolved gases such $SO_2$, $O_3$ and $NO_2$ during the sampling at pH=9 with Cellulose filters using bicarbonate buffer. In particular, Cr(VI) chemistry on sampling filters is simulated. Initial concentrations for conceptual model validation are obtained from field measurements of atmospheric Cr performed in New Jersey (Amouei Torkmahalleh et al., 2013). In the present study, it was found that the dominant form of Cr(VI) in the solution is $CrO_4^{2-}$(Figure 1). Moreover, $CrO_4^{2-}$ was produced by the dissolution of $Na_2CrO_4$(s) and $K_2CrO_4$(s) available in the solid core, but a considerable portion of the $CrO_4^{2-}$ precipitated as $(NH_4)_2CrO_4$(s), $CaCrO_4$(s), $BaCrO_4$(s) and $PbCrO_4$(s). The $HCrO_4^-$ and $Cr_2O_7^{2-}$ fractions of soluble Cr (VI) in the aqueous layer were significantly low which is almost negligible. In general, conversion of Cr(VI) to Cr(IIII) was higher than the conversion of Cr(III) to Cr(VI). The major reductant was found to be $VO^{2+}$ (Figure 2). The simulation results agree with the field measurements near Cr industries. The results of this study provide new insights into the solubility of Cr(VI) species, and helps to better design sampling filters and analytical methods to quantify Cr species.

# 512838 Has Syrian War Impacted the Atmospheric Pollutants?

**Abstract:**

Vertical Density Column (concentration) of $NO_2$ over the atmosphere of the Syrian Arab Republic during the 2005-2016 years period was studied to identify the impact of the Syrian War on its atmosphere. The concentration values of $NO_2$ were retrieved from the satellite data collected on the Ozone Monitoring Instrument on the board of Aura Satellite and analyzed through NCAR Command Language. This study investigates the impact of the war in the Syrian Arab Republic on $NO_2$, VOCs and $O_3$ concentrations before and during the war. It was observed that during the period of 2005-2011, average concentration of $NO_2$ experienced a gradually increasing trend due to the rise of the Syrian population, vehicle's use and industrial growth. Particularly, an acceptable correlation was observed between the oil production in Syria and the $NO_2$ column. From 2011 to 2012 when was the starting period of the war (street protests), the $NO_2$ column was constant. The war impact on the amount of $NO_2$ can be noticed by a sharp decrease of nitrogen dioxide column starting from 2012, the turning point of the Syrian war when the vigorous armed interactions have begun. The main effect of war can be explained by the considerable reduction of the population caused by the refugee migration, disruption of power and chemical plants and explosions due to the major chemical attacks. We are now currently analyzing the impact of the war on VOCs and ozone concentrations to better understand the influence of the war on the atmospheric chemistry of Syria. The results will be presented to AIChE 2018 Spring meeting.

# 513586 Environmental Risk Analysis of Accidental Releases from Onshore Oil Pipelines

**Abstract:**

Accidental releases of oil and oil products will cause extensive damage to environment, including soil and water contaminations, if timely and effective measures are not available. Predicting the consequences of spilled oil is of significant importance for emergency management. Although software for risk assessment of gas pipelines is very popular, few are available for hazardous liquid pipelines, due to the difference in behaviors of accidental releases of gases and liquids in the same situation. The major differences are that the spread of released oil is mainly affected by the topography of the land and may result in pollution of soil or waterways, while gas pipeline failure may form gas clouds or explosions and merely pose environmental pollution problems. An integrated model was developed in order to analysis the environmental consequences of spills from oil pipelines. Specific environmental risk indexes were defined, expressing the risk of soil and groundwater contamination, both in physical and economic terms. A case-study is presented and discussed to illustrate the features of the methodology. The results confirmed that the proposed model may be considered an important tool within a comprehensive approach to the management of risk related to onshore oil pipelines.

## 512019 Design, Development and Demonstration of a Tail-End Acetylene Hydrogenation Catalyst

**Abstract:**

Steam cracking of hydrocarbon feeds is the preferred commercial route for the production of ethylene. The yield of by-product acetylene is typically between 0.5 and 2.5 kg per metric tonne of ethylene produced. The modern specification for acetylene in ethylene product is less than 1 ppm, due to downstream process requirements. However conventional distillation cannot separate acetylene from ethylene, therefore selective hydrogenation is practiced in most plants.

Palladium catalysts have been used for the hydrogenation of acetylene for a number of years. In an industrial plant with a tail-end configuration the ethylene purification step is down-stream of the de-ethanizer and usually consists of multiple catalyst beds in order to manage the heat of reaction and maximize ethylene yield. Within a multiple bed system, the lead bed typically removes the majority of the acetylene from the ethylene stream, up to 80%. Downstream reactors are used to lower the acetylene level to the sub ppm levels required to meet the specification. Tail-end systems control selectivity both through the selection of an appropriate catalyst, and control of over-hydrogenation through hydrogen stoichiometry and bed temperature. Efficient removal via selective hydrogenation presents some unique challenges for heterogeneous catalysts. Aside from the high activity and selectivity targets associated with such duties the propensity of catalysts within a tail-end selective hydrogenation system, particularly in the lead bed where the catalyst is in a hydrogen lean environment, to form green-oil is well known. Whilst the formation of green oil, a mixture of C4 to C20 unsaturated components, is well known and has been observed for a number of years, the formation mechanism is poorly understood and little studied. Formation of green-oil lowers the performance of the hydrogenation catalysts and requires frequently regeneration to remove the polymeric carbon deposits. Such regenerations are difficult to manage and require several days of down-time on the manufacturing facility.

Critical factors in maximizing activity, selectivity and controlling the rate of deactivation include formulation, metal/promoter interactions and both physical and chemical attributes of the catalyst/carrier material, coupled with reproducible and cost effective catalyst manufacturing methodologies. To understand the catalyst characteristics, and effectively design and manufacture a catalyst to overcome the challenges faced within industrial operation, innovative approaches must be taken to understand the reaction mechanisms and interactions between the components of the catalyst and the reactants.

Within this paper we will present our approach to the understanding of green-oil formation, and the strategy to translate this into improved catalyst formulations. Through advanced characterization, coupled with state of the art testing facilities, we can apply the fundamental understanding of the process requirements we have developed into industrial catalysis, both through the launch of new products and by working closely with industrial customers to enable the optimization of their plant operations. Within our presentation we will demonstrate the methodologies weâ€™ve applied and how these have advanced our catalyst offerings for selective hydrogenation. We will also highlight the improvements to ethylene yield and cycle length that have been attainable on an industrial plant by applying the knowledge gained through extensive testing of our formulations under different process conditions at laboratory scale.

# 512679 Comprehensive Improvement of the Performances of Light Cuts Hydrogenation Units

**Abstract:**

In order to help petrochemical plant managers to increase the availability and reliability of their units, Axens developed a global offer including adsorbents and high performances catalysts to increase the selectivity and the cycle length of light cuts selective hydrogenation units, in the continuity of its very successful and well-known AxTrap, AxSorb and LD series.

Axens has developed advanced purification solutions to cope with all contaminants present in C3 and C4 cuts, especially when processing ROG streams, as well as catalytic solutions that allow improving the performances of the unit. The most recent sacrificial AxTrap and regenerable AxSorb adsorbents in combination with the well-known LD catalysts family will ensure a complete optimization of the purification and catalytic systems.

LD catalysts have demonstrated a significant improvement compared to the competition: either on selectivity and cycle lengths basis. Industrial case studies will illustrate the significant economic gain while keeping stable activity, excellent regenerability and poisons high tolerance of the overall system.

# 513066 Evaluating the Impact of Higher Emissivity Technologies on Steam Cracking Furnaces, By Means of COILSIM1D Simulations

**Abstract:**

**Evaluating the impact of higher emissivity technologies on steam cracking furnaces, by means of COILSIM1D simulations**

**Session:** 30th Ethylene Producers' Conference

**Authors:** Andrés E. Muñoz G.[1], David J. Brown[1]

1. AVGI bvba

**Abstract:**

Olefin production by thermal cracking is a mature process and so major innovations are comparatively rare. However, because of both the huge scale and the steady growth of the ethylene industry worldwide, several technologies are being developed that reduce energy consumption of cracking furnaces. For some years already, one of the fields of focus for improvement has been the enhancement of emissivity of the furnace refractory lining (the 'hot face') and the outside surface of the radiant coils by applying coatings.

The principle is simple: Higher emissivity in the radiant box ensures more efficient heat transfer to the process gas and, therefore, less energy (expressed, for example, as fuel flow rate) is required in order to maintain the same cracking severity, when compared to a normal emissivity furnace. However, because of a lower amount of flue gas being generated, the arch temperature will be lower (see Figure 1) and the heat recovery taking place in the convection section will be less, offsetting the beneficial effect originally expected from the increase in radiant efficiency. Consequently, an adequate balance of the overall effect of enhanced emissivity is of the utmost importance when considering the application of this technique.

This paper illustrates the impact of increasing emissivity in the radiant section, by means of COILSIM1D simulations. COILSIM1D can model accurately all the interacting and competing effects within an industrial furnace. Combining a detailed microkinetic model to describe the chemistry taking place in the reactor, with a coupled radiant box – convection section furnace model, the global effect of increasing emissivity can be studied.

The conclusion of the study is that high-emissivity coatings can be very beneficial but that a proper evaluation must made, case-by-case, to choose the right candidates.

**Figure 1:** Arch and Coil Outlet Temperature as a function of emissivity, for a constant firebox fired duty

## 513731 Process Optimization and Iow Implementation for Preventing Loss of Containment in Gas Processing Plants (amine treatment)

**Abstract:**

Gas treating plants have to increase their processing capacity, because of the new customer requirements. In order to be able to provide the amount of sweet gas required by the market, process conditions were altered without any prior analysis. The sour gas stream enter the absorber column (AC), in past years, the contaminants like $H_2S$ and $CO_2$, were adequately controlled, but, nowadays, due to some composition changes in the feed stream, contaminants concentrations has increase. So, with more contaminants on the stream, more lean-amine is required to remove them. The lean-amine is obtained thru a regeneration process which implies heating the rich-amina stream from the AC, using steam as a heat source. This process is carry out at the regeneration column (RC). In order to supply a more lean-amine, the reboiler and RC temperature should be increased, this causes the amine to degrade, and this degradation mechanism produces heat-stable amine salts (HSAS), this conditions increases corrosion rates and wall thickness reduction, which leads to a loss of containment. An amine substitution program was implemented, also an IOW program is proposed to be implemented, mainly for RC temperature control, this IOW should be implement together with a quality/process window for $H_2S$ and $CO_2$ concentration control. The benefits for these alternatives bring an increase of sweet gas production, from 208 MMSCFD to 218 MMSCFD; energy savings (MMBTU) form USD 90,963 to USD 86,724. In addition, loss of containment reduction is expected, which influences safety, reliability and business continuity.

## 513680 A New Model to Predict MEA Loss through Particulate Phase (Aerosol)Â  in Post Combustion CO2 Capture Plants Using Aspen Plus

**Abstract:**

The Monoethanolamine (MEA) aqueous solution is widely used as a solvent for absorbing CO2. One of the disadvantages of MEA is that exceedingly high amount of energy is required which causes loss of overall efficiency of power plants. Moreover, regardless of solvent mechanical loss, the solvent loss due to evaporation and particulate phase emissions is another issue that results in negative environmental effects and increase of the total operational cost. Experimental studies on PM measurements in PCCC columns are scarce in the literature, necessitating developments of simulation and modeling studies in this field. However, prediction of particle formation in PCCC columns is a complex task requiring modeling of several steps including gas-phase supersaturation, particulate matter (PM) formation (nucleation) and PM dynamics (growth and loss). Preexisting PM introduced by combustion fumes into PCCC facilitates formation of new particles as a result of gas phase supersaturation. Hence, developing a simulation tool to understand the key parameters affecting the supersaturation of the gas phase in PCCC columns is of paramount importance. Few attempts were made to setup a simulation tool using Aspen Plus simulator to predict supersaturation of gas streams in PCCC columns in the absence of particulate phase [14], and in the presence of combustion generated PM in the flue gas.[15] However, such simulation models still need to be improved to better understand the interactions between the particulate phase and other phases inside the column. In the current study a new conceptual model based on the film theory was developed to study interaction between the particulate matter and the gas and the solvent flows inside an absorbing column. The new model gives opportunity to study aerosol emission exiting from the absorber column, i.e. its flowrate and composition, and factors that have influence on it. Aspen Plus V9.0 software was used to model the system. An absorber column, and solvent and flue gas inlet parameters were taken from a pilot plant study at the University of Texas at Austin. The current model enable us to distinguish between the MEA loss through particulate phase and the gas phase. The results show an increase in particle size at the top of the column which could be due to the condensation of materials on the surface of the existing particles inside the column. The glue gas becomes supersaturated at the top of the column due to the rapid cooling. Further results will be presented during the AIChE meeting.

# Reliable Experimental Technique for the Identification of Foam Creators from Industrial Lean Amine Solutions

Emad Alhseinat[a*], Mohammed Keewan[b], Fawzi Banat[b]

[a] *Department of Chemical Engineering, Khalifa University of Science and Technology, P.O. Box 127788, Abu Dhabi, United Arab Emirates*
[b] *Department of Chemical Engineering, Khalifa University of Science and Technology, Petroleum Institute, P.O. Box 2533, Abu Dhabi, United Arab Emirates*

** Corresponding author: PO Box 127788, Abu Dhabi, UAE*
*E-mail address: emad.alhseinat@kustar.ac.ae*

## Abstract

In this paper, a new technique based on the concept of foam fractionation is developed for the identification of foam creators in alkanolamine gas sweetening solvents. The developed foaming technique was used to identify the foam creators existing in industrial amine solutions (lean amine from the regeneration column outlet) obtained from running a gas sweetening plant. The developed approach in this technique is based on inducing a lean amine sample to foam, allowing three samples to be collected: 1) A sample from the lean amine before foam (main amine sample) 2) A sample from the collected foam (foamate) 3) A sample from the remaining amine after collecting the foam. Direct sample analysis time of flight (AxION® 2 TOF) instruments were used to analyze the collected samples. A comparison was made between the three spectra obtained from a mass spectrometer for the three samples. The compounds that increased in concentration in the foamate sample and decreased in the remaining amine in the foaming column are believed to be responsible for generation of the foam. The obtained mass/charge from the mass spectrometer can be used to identify these compounds. Identification of foaming creators in industrial lean amine samples is provided as a proof of the capabilities of the proposed method. The results of this study introduce the foaming technique as a promising laboratory scale technique for the identification of foam creators in amine absorption solvents.

**Keywords**: Amine foaming; MDEA; foam fractionation; foam creators

## Introduction

Alkanolamine absorption processes are widely used to remove acid gases i.e. $CO_2$ and $H_2S$ from Natural gas and flue gases Mokhatab;, POE; et al. (2012), Alhseinat, Pal et al. (2014). Foaming is one of the most severe operational problem in acid gas absorption process. Foaming has severe impact on plant operation, causing excessive loss of absorption solvents, premature flooding, reduction in plant throughput, and off-specification of products Thitakamol and Veawab (2009). Foaming problem can be induced by several contaminants including fine particulate, condensed liquid hydrocarbon, heat stable salts (HSS) and amine degradation products (Alhseinat, Pal et al. 2014, Alhseinat, Pal et al. 2015, Alhseinat, Keewan et al. 2017). Due to the closed loop nature of amine processes, foam creators can accumulate to reach their effective concentrations without being noticed by the unit operators (Pauley 1991, Chen, Freeman et al. 2011). Unfortunately, there is no method for foam source identification; despite the fact that identification of foam sources is key to troubleshooting the problem of amine foaming (Dhafeeri 2007).

Adsorptive bubble separation is used for partially separating or concentrating dissolved material by adsorption at the surface of bubbles based on the difference of surface activity (Darton, Supino et al.),(Swain 2005). The compounds responsible for foaming in gas sweetening unit tend to accumulate in the foam itself, it is therefore possible to concentrate them by inducing the amine to foam in an experimental foam fractionation setup; this ultimately facilitates their identification and removal.

Phul et.al (Phul 2003) proposed a method and apparatus for removing foaming contaminates from solvents using the concept of foam fractionation. The proposed apparatus was designed and manufactured to clean all circulated lean amine or at least a significant portion of it in order to be effective in preventing foaming at amine gas sweetening processes. Indeed, nothing is mentioned in the work of Phul et.al (Phul 2003) about the monitoring of solvent foaming behavior and no methodology is proposed for the identification of foam creators. There is a clear lack of scientific research investigating and optimizing the foam fractionation technique for amine gas sweeting solvent cleaning and the identification of foam creators. Therefore, to enhance our knowledge about the use of foam fractionation for amine solvents cleaning and foam creators' identification, a first attempt for optimizing batch foam fractionation column using laboratory synthesized amine solutions from analytical grade amine was performed and published in Journal of Natural Gas Science and Engineering (Alhseinat, Amr et al. 2015). These efforts led to the development and optimization of a foam fractionation setup for the identification of foam creators in industrial lean amine. This paper introduces for the first time, an experimental technique and approach based on foam fractionation for the identification of foam creators in industrial lean amine solution. In this study, a continuous foam fractionation set-up was designed and used to concentrate the foam creators from industrial lean amine. Direct sample analysis time of flight (AxION® 2 TOF) instruments were used to analyze the collected foam, the main amine solution and the remaining liquid following foaming. A comparison between the concertation of the species existing in the collected foam, the initial amine solution prior to foaming and the amine solution following foaming (named the remaining liquid) is the key to identifying the foam source, which will be explained in detail later in the paper. The results of this study highlight, for the first time, the ability of foam analysis to identify foam creators using a simple experimental setup. This research enhances the understanding of the contributing factors to the foaming problem and will facilitate the tracking of the sources of foaming and facilitate their alleviation or impact reduction.

## Materials and Methods

The identification and detection of foam creators was carried out for lean MDEA solution obtained from GASCO (A gas sweetening company in the United Arab Emirates). A continuous foam fractionation column was used to concentrate the foam creators existing in industrial lean amine. Direct sample analysis time of flight mass spectrometry (DSA-ToF MS) instrument was then used to detect and identify the foam creators.

### Materials

Industrial lean MDEA was obtained from GASCO, Habshan, Abu Dhabi.

### Experimental Setup

As illustrated in Figure 1, the experimental setup was composed of a foam fractionation column with a cross sectional area of 0.0022 $m^2$, and a flow meter. Nitrogen gas ($N_2$) was used to generate bubbles in the tested amine solution. A peristaltic pump was used to compensate the lean amine solution level in main column; in addition to a vacuum pump that was used to withdraw the foamate sample.

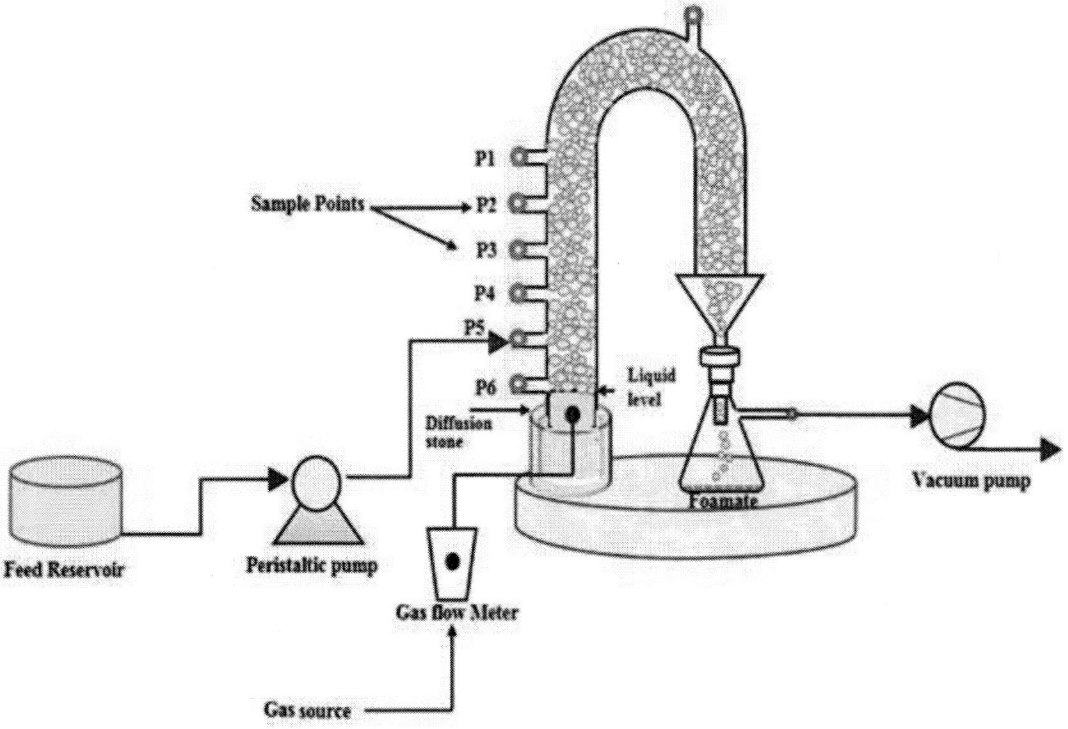


*Figure 1: Schematic diagram for Continuous foam fractionation setup*

## Experimental Procedures and Conditions

### *Foam Fractionation*

In the foam fractionation experiments, three samples where collected from each experiment and tested using Direct sample analyzer Time of Flight mass spectrometry (DSA-ToF MS) and Liquid Chromatography Time of Flight mass spectrometry (LC-ToF MS): the main solution sample (before foaming treatment): the foamate sample and the liquid sample (remaining MDEA solution in the foam fractionation column). For the foaming treatment, foam was collected until no further foam was accumulated.

Several samples were analyzed to evaluate the repeatability of the results. Selected experiments were repeated to test the reproducibility of the results. Selected foam and breaking time experiments were repeated three times. The results from the repeated experiments demonstrated almost exact readings for foam volume with less than 1% deviation and 2% deviation in the breaking time readings.

*Direct Sample Analysis (DSA) Time of Flight (ToF) Mass Spectrometry (MS) (AxION 2 ToF) Instrument*

A DSA ToF MS (AxION 2 ToF MS, Perkin Elmer) instrument was used to analyze the collected samples and to identify foam creators in lean amine samples. DSA, which is integrated with a (TOF- MS), is one of the most recent and development ambient ionization techniques; these techniques enable the investigation of a variety sample types across the molecular weight range. Accurate peak intensity can be recorded across the complete mass range, mass (m)/charge (z), by AxION 2 ToF MS (PerkinElmer) instrument; it records ions and time, which is particularly important to identify and quantify all contaminant components in the sample. The DSA LC ToF MS was used previously for the identification of MDEA degradation products with high concentrations (Pal, AbuKashabeh et al. 2014). Small quantities of prepared sample without any dilution or further treatment were sent to the DAS instrument via disposable mesh; (10 μL of prepared sample), rapid screening of samples was then conducted. The obtained full spectrum measurement captured the complete composition of every sample tested. These spectra were used to identify the foam creators collected in the samples tested. The identification of compounds in this study was based on the previous degradation studies and proposed degradation mechanisms of amine by the research group using the DSA TOF MS database provided by the supplier (Perkin Elmer).

*Sample Preparation for LC ToF MS*

5 μL from the foamate, liquid or main solution sample were added to 1500 μL of water (particular water supplied to LC ToF MS Instrument), then 10 μL of this diluted sample was introduced to the LC ToF MS.

## Results and Discussion

### Surface tension change alnog the foam fractionation column

Table 1 shows the surface tension of collected samples for industrial lean MDEA (unit 30, GASCO, Habshan). The surface tension reults along the foam fractionation column were plotted in Fig. 2. It can be observed that the the surface tension decreases with the increase in column height, leading to an increase in surfactant concentration. The concentration of compounds resposible for foaming, therefore, incresaes with foam height.

***Table 1: Surface Tension of collected samples for industrial Lean MDEA and calculated efficiency parameters***

| Collected Sample Name | Sample Distance from Bottom (cm) | Surface Tension (mN/m) | R% | Enrichment |
|---|---|---|---|---|
| **P1 Foamate** | 17.0 | 21.6 | 57.0 | 2.28 |
| **P2 Foamate** | 14.5 | 24.0 | 52.0 | 2.06 |
| **P3 Foamate** | 12.0 | 26.2 | 48.0 | 1.89 |

| | | | | |
|---|---|---|---|---|
| **P4 Foamate** | 9.50 | 31.2 | 38.0 | 1.58 |
| **P6 Foamate** | 4.50 | 34.1 | 32.0 | 1.45 |
| **Liquid Sample** | 2.50 | 50.5 | - | - |
| **Main Sample** | 0 | 49.5 | - | - |

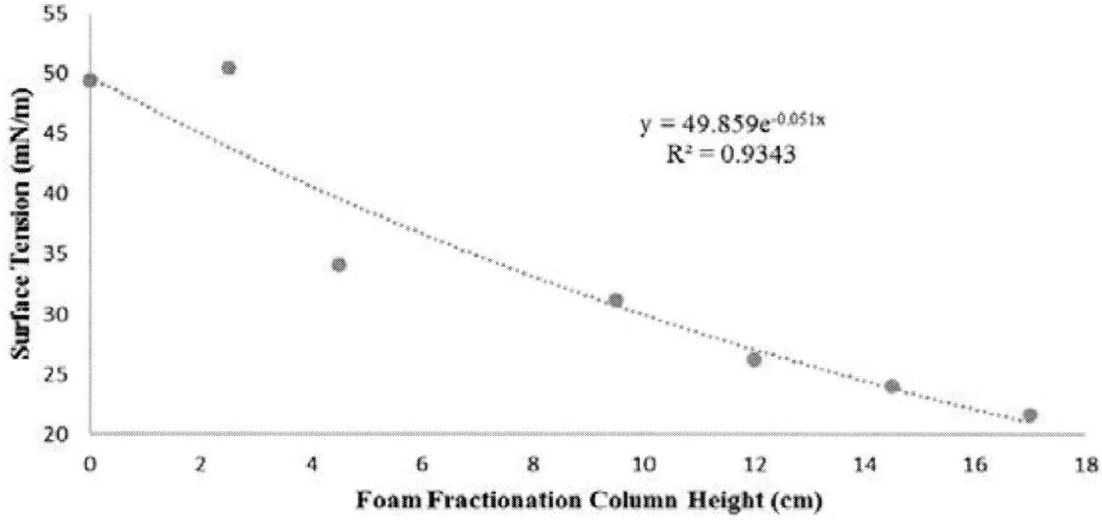


*Figure 2: Surface Tension relationship with Column Height*

## Direct Sample Analysis (DSA) Time of Flight (ToF) Mass Spectrometry (MS) Results

Figure 3 shows the DSA ToF MS analysis results for the P1 foamate sample (17 cm) and the P3 foamate sample (12 cm) obtained from lean amine (unit 30, GASCO, Habshan). As expected the absolute amplitude for MDEA, mass (m)/charge (z) (m/z) 120.1002 was the highest amplitude in all analyzed samples; the ratio between absolute amplitude for each peak to highest peak (MDEA) is the relative amplitude % values that are shown in all DSA ToF MS figures below.

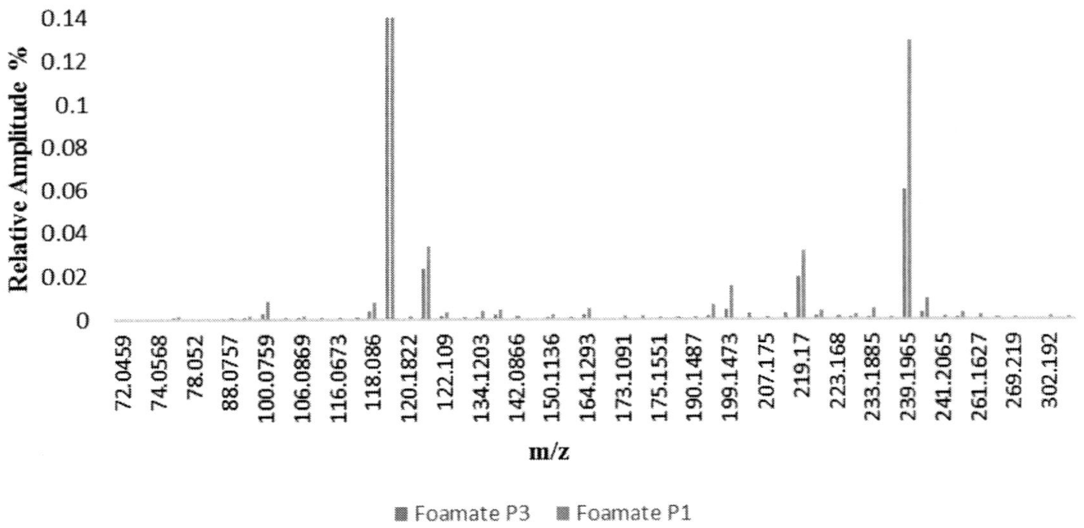


*Figure 3: DSA ToF MS analysis for P1 foamate sample and P3 foamate sample obtained from industrial lean Amine.*

As shown in Fig. 3, there are some compounds such as (m/z) N-Methylpyrrolidone ($C_5H_9NO$), m/z: 100.0759, Protonated Methyldiethanolamine m/z: 121.1044, Methyltriethanolamine (MTEA), m/z: 164.1293, 4-Hydroxy-5-(3-methylbutoxy) pentane hydrazide ($C_{10}H_{22}N_2O_3$) m/z: 219.17 and N,N,N',N'-Tetrakis(2-hydroxyethyl)-1,2-ethanediaminium (2.Methyldiethanolamine with hydrogen) $C_{10}H_{27}N_2O_4$, m/z: 239.1965 that show peaks with higher relative amplitude % in P1 foamate (17 cm) than in P3 foamate (12 cm). Which means that these compounds were concentrated in the P1 foamate sample (highest point in fractionation column)

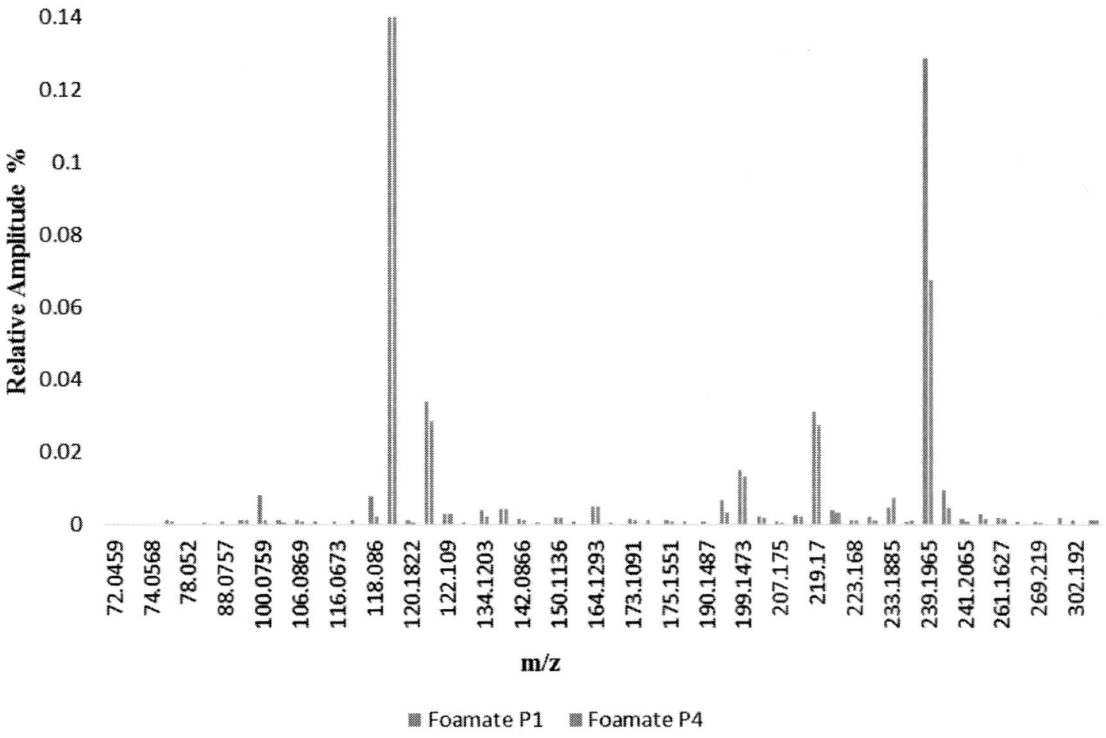


***Figure 4: DSA ToF MS analysis for P1 foamate sample and P4 foamate sample obtained from industrial lean amine.***

Figure 4 shows the relative amplitudes values of all compounds (m/z) for P1 foamate (17 cm) and P4 Foamate (9.5 cm) samples. As can be observed, most peaks for the majority of compounds (m/z) have higher relative amplitude % in P1 foamate sample than in P4 foamate sample.

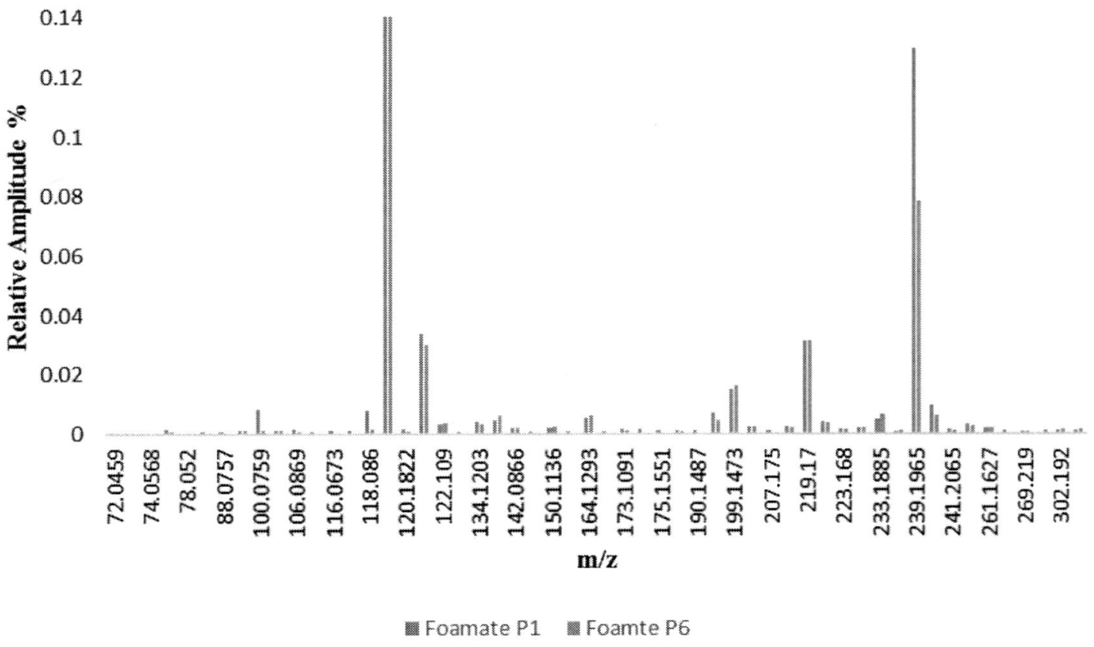


***Figure 5: DSA ToF MS analysis for P1 foamate sample and P6 foamate sample obtained from industrial lean amine.***

Similar observations were noticed in comparing relative amplitude values of all compounds (m/z) between P1 foamate (17 cm) and P6 foamate (4.5 cm), as can be seen in Fig. 5. Peaks for most compounds (m/z) have a higher relative amplitude % in P1 foamate sample (17 cm) than in P6 foamate sample (4.5 cm).

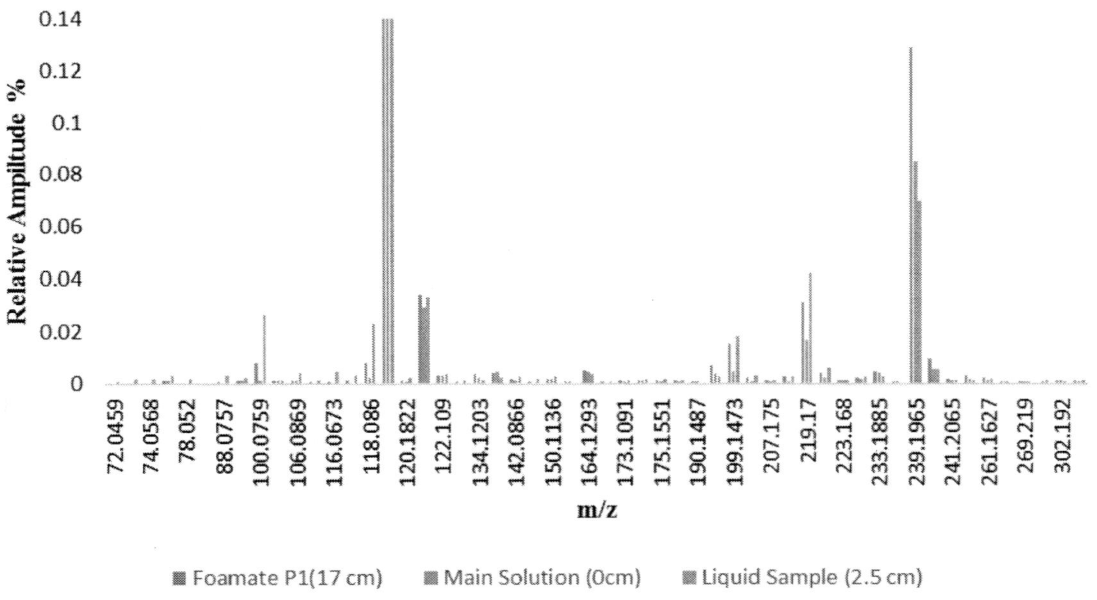


***Figure 6: DSA ToF MS analysis for P1 foamate sample, Liquid sample and Main solution sample obtained from industrial lean amine.***

Figure 6 shows a comparison between the relative amplitudes values of all compounds (m/z) that were detected in the P1 foamate sample (17 cm), main lean amine solution (0 cm) and the remaining liquid sample (2.5 cm) in the foam column. It can be seen that almost all degradation products and other contaminates have lower relative amplitudes values in the remaining liquid than the initial relative amplitudes values in the main solution. However, there are some peaks that have higher relative amplitudes values than the main solution sample. This means that in general the contaminants were concentrated in the top layers of the foamate.

## Removal Foam Creators using foaming treatment technique

Figures 7 - 10 show foaming treatment for another industrial lean amine sample. In this experiment, foam was collected until no more foam was observed.

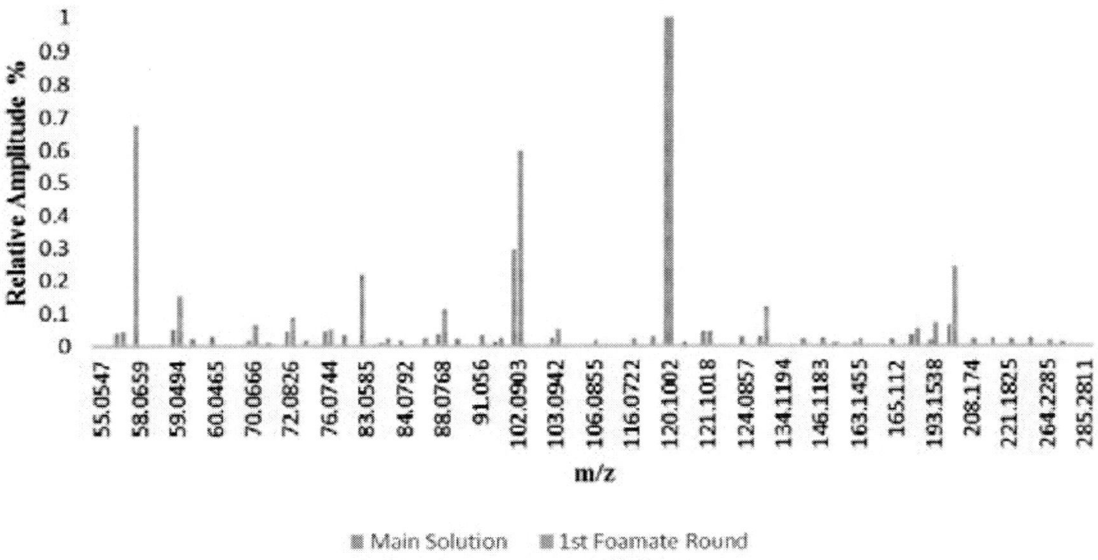


*Figure 7: DSA ToF MS analysis for main solution sample, foamate sample and liquid sample obtained from first foaming treatment round of industrial lean amine*

Figurer 7 shows the relative amplitudes values of all compounds (m/z) that were detected from the first foaming round. It can be seen that there are peaks for some compounds (m/z) that have higher relative amplitude % in the foamate sample than in main solution sample; this indicates that the following compounds: (m/z) acetate, m/z: 59.0494, dimethylacetamide, m/z: 88.07, Dimethylethanolamine, m/z: 102.0903, Diethanolamine (DEA), m/z:106.0855, Tris(hydroxyethyl)ethylene diamine (THEED), m/z: 193.1538, unknown compound m/z 203 and Methanamine,N,N-di(2-trimethylsilyloxyethyl), m/z: 264.2285 were partly separated with the first foamate sample.

Figure 8 shows the relative amplitude values for all compounds (m/z) that were detected from the second round of foaming. It can be observed in Figure 8 that the detected relative amplitude for the compounds mentioned above that were detected in the first round of foaming, decreased in the main solution and increased in the foamate sample. This indicates that further separation of these compounds is occurring. Interestingly, few compounds i.e. Diethanolamine (DEA), m/z: 106.0855, Tris(hydroxyethyl)ethylene diamine (THEED), m/z: 193.1538 and Methanamine,N,N-di(2-trimethylsilyloxyethyl), m/z: 264.2285 were detected in the first treatment round have not been detected in the second round. This suggests that the concentration of these compounds either decreased below the detection limit of the DSA-MS or they were completely removed from the solution in the first foaming round.

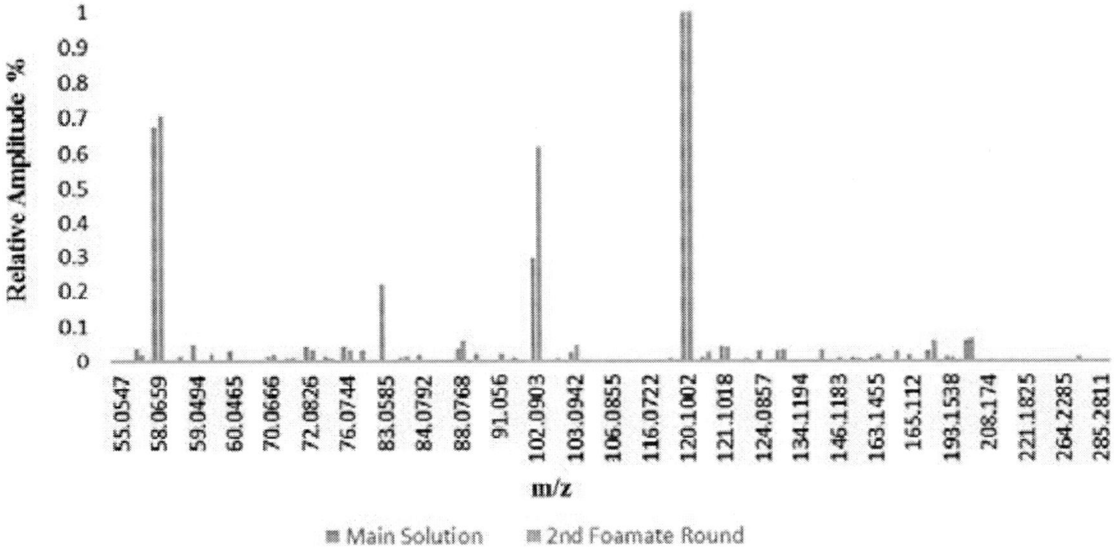


*Figure 8: DSA ToF MS analysis for main solution sample and foamate sample obtained from second foaming treatment round of industrial lean amine.*

Figure 9 displays the relative amplitude values of all compounds (m/z) that were detected from the third round of foaming. It can be observed from Figure 9 that the detected relative amplitude for all of the detected compounds in the first sample decreased in the main solution and increased in the foamate sample. It can be noticed that (THEED), m/z: 193.1538 was not detected in the third foaming round, which indicates its complete removal in the first two rounds.

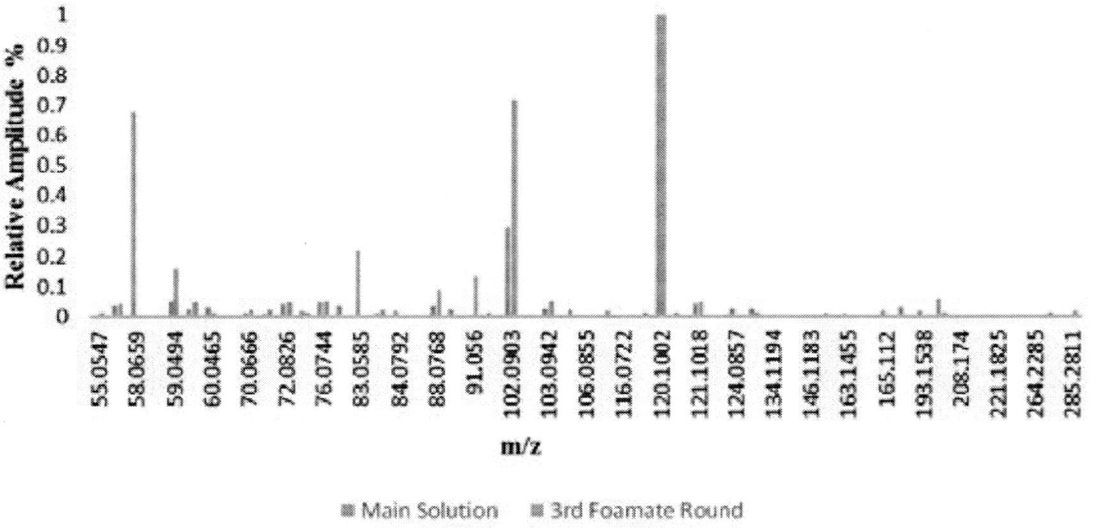


*Figure 9: DSA ToF MS analysis for main solution sample, foamate sample and liquid sample obtained from third foaming treatment round of industrial lean amine*

Figure 10 shows a comparison between the relative amplitude values of all compounds (m/z) that were detected in the main lean amine solution (before foaming treatment) and the remaining liquid (remaining lean amine) in the foam column after the foam treatment. It can be observed that almost all degradation products and other contaminants have lower relative amplitude values in the remaining liquid than the initial relative amplitudes values in the main solution. This provides clear evidence that the concentrations of these compounds are decreased due to the foaming treatment. We can therefore say that the foaming treatment facilitates the removal and identification of the foaming creators in real industrial lean amine.

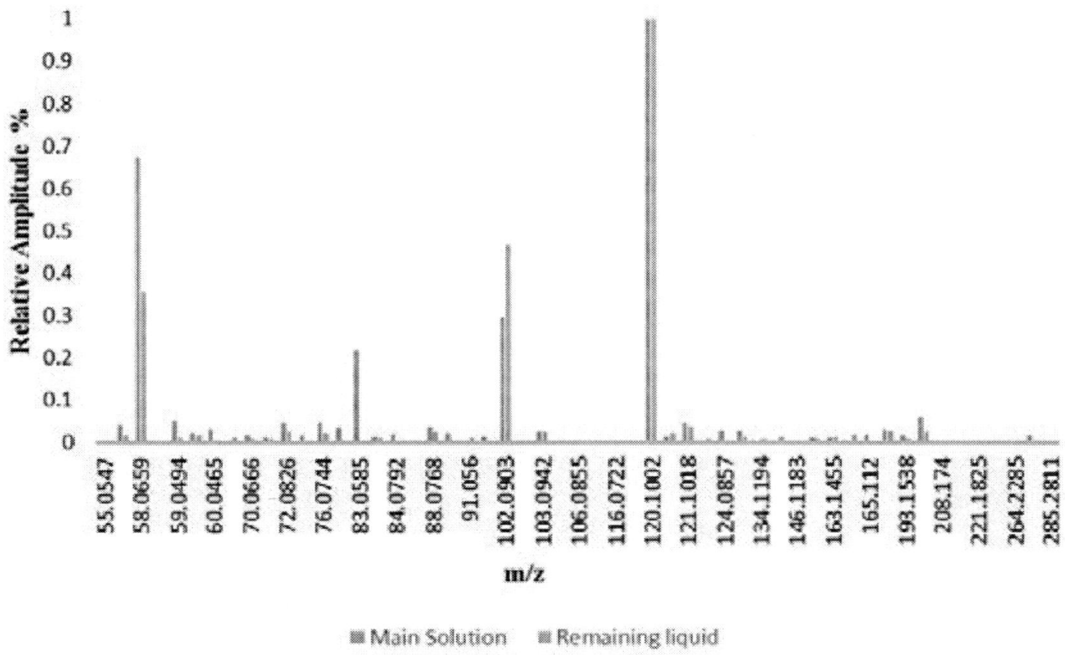


*Figure 10: DSA ToF MS analysis for initial main solution sample and remaining liquid sample obtained from multifoaming treatment of industrial lean amine.*

## Conclusion

Fractional foaming was used for the identification and the removal of foam creators and other contaminants from the industrial lean amine sample. The compounds (m/z) acetate, m/z: 59.0494, Dimethylacetamide, m/z: 88.07, Dimethylethanolamine, m/z: 102.0903, Diethanolamine, m/z: 106.0855 (DEA), Tris(hydroxyethyl)ethylenediamine (THEED), m/z: 193.1538, and Methanamine,N,N-di(2-trimethylsilyloxyethyl), m/z: 264.2285 that were attached to the foam were identified as possible foam creators. Some contaminants were removed from the first round of foaming while others required more rounds for their complete removal. Fractional foaming was proven as a potential technique for the identification of foam creators with very low concentrations that are undetectable using traditional methods.

## Acknowledgments

The authors would like to thank the Petroleum Institute Gas Processing and Materials Science Research Center, Abu Dhabi for funding the project (GRC006). Also the authors would like to thank Dr. Doug Aitken from Universidad de Concepcion for editing the paper.

## References

Alhseinat, E., M. Amr, R. Jumah and F. Banat (2015). "Removal of MDEA foam creators using foam fractionation: Parametric study coupled with foam characterization." Journal of Natural Gas Science and Engineering **26**: 502-509.

Alhseinat, E., M. Keewan and F. Banat (2017). "Impact of dissolved and undissolved organics on foaming of industrial amine." International Journal of Greenhouse Gas Control **60**: 156-161.

Alhseinat, E., P. Pal, A. Ganesan and F. Banat (2015). "Effect of MDEA degradation products on foaming behavior and physical properties of aqueous MDEA solutions." International Journal of Greenhouse Gas Control **37**(0): 280-286.

Alhseinat, E., P. Pal, M. Keewan and F. Banat (2014). "Foaming study combined with physical characterization of aqueous MDEA gas sweetening solutions." Journal of Natural Gas Science and Engineering **17**(0): 49-57.

Chen, X., S. A. Freeman and G. T. Rochelle (2011). "Foaming of aqueous piperazine and monoethanolamine for CO2 capture." International Journal of Greenhouse Gas Control **5**(2): 381-386.

Darton, R., S. Supino and K. Sweeting Development of a Multistageed Foam Fractionation Column. Oxford, England.

Dhafeeri, M. A. A. (2007). Identifying sources key to detailed troubleshooting of amine foaming, Oil and Gas Journal.

Mokhatab;, S., W. POE; and J. Speight (2012). Handbook of Natural Gas Transmission and Processing,. USA, Elsevier.

Pal, P., A. AbuKashabeh, S. Al-Asheh and F. Banat (2014). "Accumulation of heat stable salts and degraded products during thermal degradation of aqueous methyldiethanolamine (MDEA) using microwave digester and high pressure reactor." Journal of Natural Gas Science and Engineering **21**: 1043-1047.

Pauley, C. R. (1991). "Face the facts about amine foaming." Chemical Engineering Progress **87**(7): 33-38.

Phul, S. A. v. (2003). METHOD AND APPARATUS FOR REMOVING FOAMING CONTAMINANTS FROM HYDROCARBON PROCESSING SOLVENTS. U. S. Patent. USA.

Swain, M. (2005). Foam Fractionation Risers, Part II Examinations. Foam Fractionation Risers. Oxford.

Thitakamol, B. and A. Veawab (2009). "Foaming model for CO2 absorption process using aqueous monoethanolamine solutions." Colloids and Surfaces A: Physicochemical and Engineering Aspects **349**(1–3): 125-136.

# 511847 Purification of Styrene from the Styrene/Ethylbenzene Mixture with a Small Amount of the Mixed Xylene By Stripping Crystallization

**Abstract:**

Stripping crystallization (SC) is a new separation technique which combines distillation and crystallization. Basically, SC is operated at a triple-point condition, in which the liquid is simultaneously vaporized and crystallized due to the three-phase equilibrium. In essence, the process is continued until the liquid phase is completely eliminated and only the pure solid crystals remain in the feed. Thus, crystal washing is not required since only pure solid crystals remain in the feed and no impurities are adhered on the crystal surfaces at the end of SC. In the previous research, SC has been successfully applied to separate p-xylene from the mixed xylenes by Shiau et al. (2005, 2008, 2013).

SC is introduced in this work to purify styrene (ST) from the multi-component system consisting of ST, ethylbenzene (EB), p-xylene (PX), m-xylene (MX) and o-xylene (OX). Styrene is one of the most important aromatic compounds and is used extensively in the manufacture of polystyrene. Due to the close boiling points of ST, EB, PX, MX and OX, it is rather complicated and energy-intensive to separate them by conventional distillation. A thermodynamic model is developed to determine the three-phase equilibrium conditions and direct the SC experiments. The experiments show that SC can be effectively applied to purify ST from the mixture. However, as EB is more easily incorporated into the ST crystal lattice compared to PX, MX and OX, it becomes more difficult to further purify ST for a higher concentration of EB in the mixture.

References:

1. Shiau, L. D., Wen, C. C. and Lin, B. S. "Separation and purification of p-xylene from the mixture of m-xylene and p-xylene by distillative freezing", *Ind. Eng. Chem. Res*, **44(7)**, p2258-2265, 2005.
2. Shiau, L. D., Wen, C. C. and Lin, B. S. "Separation of p-xylene from the multicomponent xylene system by stripping crystallization", *AIChE Journal*, **54(1)**, p337-342, 2008.
3. Shiau, L. D.; Liu, K. F. "Investigations into the effects of the cooling rate on stripping crystallization", *Ind. Eng. Chem. Res*, **52**, p1716-1722, 2013.

## 513588 Rapid Catalyst Deactivation and Lifetime Determination

**Abstract:**

Hydrotreating process is the key to refining industry to meet fuel specifications and environmental regulations. Activity and stability of hydrotreating catalyst is extremely important in hydrotreating, as it determines how long the reactor can be operated before the catalyst needs to be replaced. However, the current methods for evaluating hydrotreating catalyst are very time consuming, not cost-effective, and usually performed under noncommercial process conditions. This paper reported a novel method for rapid and efficient way to estimate hydrotreating catalyst lifetime to capitalize on industryâ€™s need.

A four-place catalyst tester was used to test the activity of commercial hydrotreating catalyst in a short time by accelerated deactivation of the catalyst using spiking agents. Accelerated hydrotreating catalyst deactivation was achieved using spiking agents of DBT and carbazole in n-dodecane, ultralow sulfur diesel and off-road diesel: the catalyst deactivation for HDN activity was 16.2% after 40-hour run. A novel methodology for catalyst lifetime determination was obtained and illustrated as the following sequences.

# Study on the Adsorptive Removal of Sulfur Compounds Present in Fuel

*Biswajit Saha, Anshu . and Sonali Sengupta,*

Saha.biswajit54@gmail.com, akmsingh02@gmail.com, sonalis.iitkgp@gmail.com,

*Chemical Engineering, IIT Kharagpur, Kharagpur, India*

## Abstract

The present work investigates the performance of flyash, coal dust, bentonite, laterite and sodium zeolite as adsorbents for desulfurization of synthetic fuel by batch adsorption experiment at $50^0C$ and under atmospheric pressure. A series of synthetic fuel samples containing known concentrations of thiophenol, benzothiophene and di-benzothiophene, as singular and multicomponent mixtures in isooctane at definite concentrations were prepared. The performance of coal dust was found to be the best among all other adsorbents mentioned, in removal of both single and multi sulfur compounds from synthetic fuel. The adsorbents were characterized by BET surface area analyzer and XRF. The effect of parameters such as, stirrer speed, adsorbent particle size, temperature, adsorbent amount, initial sulfur concentration and contact time between adsorbate and adsorbent were investigated for desulfurization of synthetic fuel using coal dust as adsorbent. The sulfur adsorption equilibrium with this adsorbent was fitted into Langmuir, Freundlich, BET, Halsay and Tempkin isotherms.

## Introduction

In last twenty five years, the energy usage of world has been increased to an average of 404 quadrillion per annum, where more than 40% of this energy came from petroleum [1,2]. Gasoline, jet fuel and diesel are the three major types of transportation fuels which contain significant amount of sulfur and combustion of those fuels result in the production of sulfur-oxide gases which cause severe harm in human health and environment[3,4]. In India, the existing two tier system for liquid fuel quality called Bharat Stage specification had introduced the BS(IV) grade diesel in since April 2010, which required a reduction of organic sulfur content from 350 ppm level to 50 ppm [5]. The conventional refinery-practiced desulfurization technology is hydrodesulfurization (HDS), which is a high severity process, hence required a high investment cost. Moreover, this process significantly reduces octane number by saturating alkenes and arenes by hydrogen used at high temperature and pressure [6]. Therefore, there is an urgent need to find out new and alternative pathways for desulfurization which would be cost effective, efficient and can meet the environmental regulations. Adsorptive desulfurization is one of the alternatives to HDS used for production of ultra clean fuel. Adsorption is a mass transfer process where the molecules in free phase become bound to a surface by intermolecular forces [7]. This process would be accomplished relatively at lower temperature and pressure. Several research works have been done on adsorptive desulfurization using various adsorbents. Shi et al[8] studied the adsorptive desulfurization of di-benzothiophene with 95% removal by using mesoporous carbon. Adeyi et al[7] reported the adsorption efficiency of manganese-di-oxide and zinc-di-oxide on desulfurization of crude oil, where manganesedioxide showed good result compared to zincdioxide with 0.816gm of sulfur adsorption per gram of adsorbent. Das gupta et al [5] showed the adsorption efficiency of nickel based adsorbents(NiMCM-41) on desulfurization of diesel oil by bringing down the sulfur concentration from 450 ppm to 50 ppm. Adsorptive desulfurization of kerosene and diesel oil by using zinc impregnated montmorollonite clay as adsorbent has been reported which removed 76 and 77% sulfur from kerosene and diesel oil respectively [9]. The adsorption efficiency of nano copper oxide on diesel fuel was reported to be 70% by Khodadadi et al[10]. Kumar et al[11] studied the adsorptive desulfurization of di-benzothiophene in isooctane phase by using zirconia based adsorbents with 53mg/g to 60 mg/g sulfur adsorption.

In the present study, the performance of flyash, coal dust, bentonite, laterite and sodium zeolite as adsorbents for desulfurization of synthetic fuel by batch process has been studied.. The effect of process variables such as time, temperature, initial sulfur concentration, stirrer speed, adsorbent amount and adsorbent particle size on the efficiency of adsorptive desulfurization has been studied with best adsorbent. Finally Langmuir, Freundlich, BET, Halsay and Tempkin adsorption isotherm models were fitted with adsorption equilibrium data.

## Materials and Method

### Materials

Bentonite powder and thiophenol were purchased from Loba Chemie Pvt Ltd, Mumbai, India. Benzothiophene was procured from Himedia Lab. Pvt Ltd, Mumbai, India. Dibenzothiophene was purchased from Merck Schuchardt OHG, Germany. Fly ash was obtained from Kolaghat thermal power plant, India. Laterite soil was taken from Midnapur district, West Bengal, India. Sodium zeolite was purchased from SRL Pvt Ltd, Kolkata, India. Coal was obtained from Jharkhand Coal Suppliers, India.

### Model fuel preparation

Four types of Model fuels were prepared by adding sulfur compounds such as, thiophenol, benzothiophene and dibenzothiophene in isooctane at different concentrations. The compositions of model fuels are as follows; Model Fuel 1 [MF-1]: Thiophenol (300 ppm) in iso-octane, Model Fuel 2 [MF-2]: Benzothiophene (300 ppm) in iso-octane, Model Fuel 3 [MF-3]: Di-benzothiophene (300 ppm) in iso-octane, Model Fuel 4 [MF-4]: Mixture of thiophenol (300 ppm), benzothiophene (300 ppm) and di-benzothiophene (300 ppm) in iso-octane with total sulfur concentration of 900 ppm,

### Batch adsorption experiment

A typical adsorption experiment was conducted using 40 ml model fuel and 0.27 gm of adsorbent, taken in a three necked flask housed in a water bath whose temperature was maintained at $50^0$C with a digital temperature controller cum indicator with an accuracy of $\pm 1^0$C. The experiment was conducted for 2h under stirring at 1000 rpm and samples were withdrawn at every 10 mins intervals.The samples were filtered to remove any trace of adsorbent particle before they were analyzed in HPLC [Perkin Elmer, Series 200] with reversed phase Agilent SB C-18 column and a Perkin Elmer Series 200 UV/VIS detector set at 254 nm. The mobile phase used was 90% methanol in water.

### Adsorbent Characterization

Adsorbents were characterised by BET surface area analyser, (Model: AS1 MP/Chemi-LP, USA) instrument, X-ray fluorescence spectroscope (XRF), (Model Panalytical, Axios, Netherland), The sulfur adsorption per unit mass of the adsorbent(mg/g) was calculated using equation 1 and 2, as,

$$q_e = \frac{(C_i - C_e)V}{W} \qquad \text{.........................(1)}$$

$$q_t = \frac{(C_i - C_t)V}{W} \qquad \text{.........................(2)}$$

## Results and Discussion

### Adsorptive desulfurization of MF-1, MF-2, MF- 3 and MF- 4 using different adsorbents

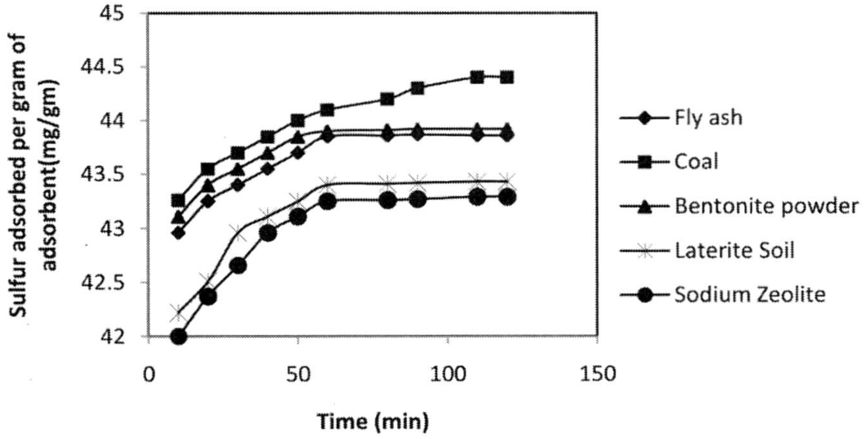


**Figure. 1A Adsorptive desulfurization of MF 1**

**(Thiophenol:-300 ppm; Adsorbent particle size:- 0.052 mm; Temperature:-50$^0$C; Adsorbent amount:-6.75gm/L of model fuel, Stirrer speed:- 1000 rpm)**

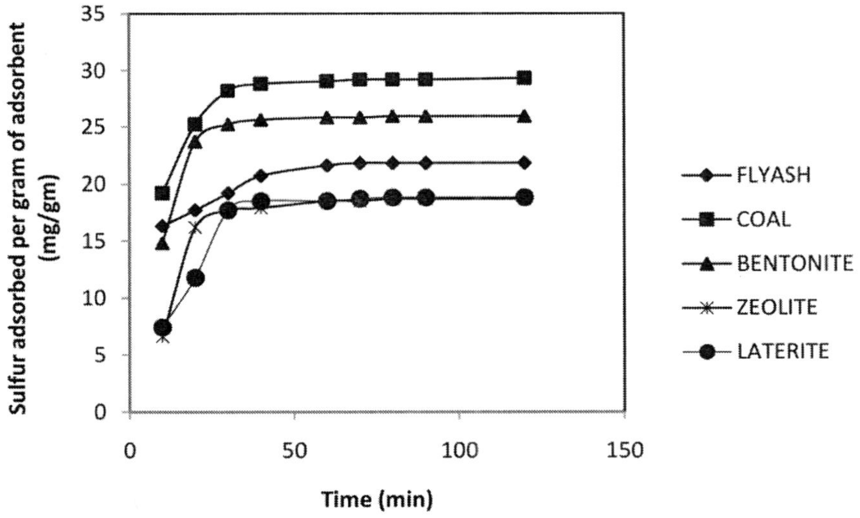


**Figure. 1B Adsorptive desulfurization of MF 2**

**(Benzothiophene:-300 ppm; Adsorbent particle size:- 0.052mm; Temperature:-50$^0$C; Adsorbent amount:-6.75gm/L of model fuel, Stirrer speed:- 1000 rpm)**

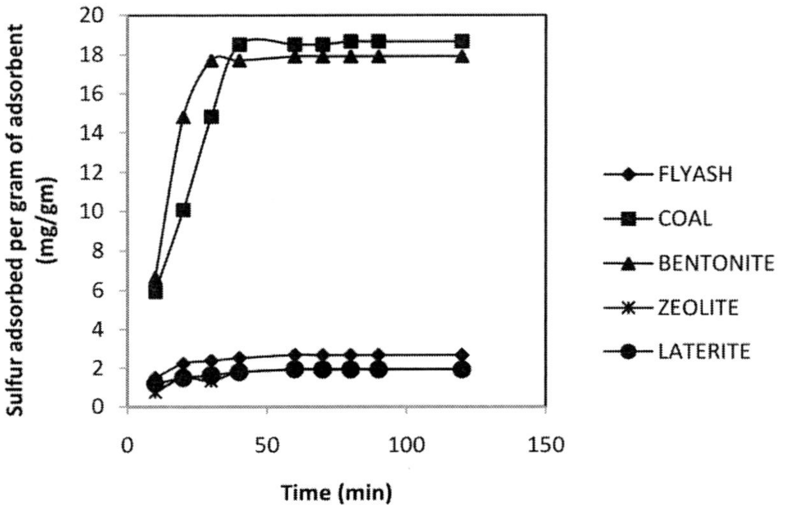


**Figure. 1C Adsorptive desulfurization of MF 3**

**(Di-benzothiophene:-300 ppm; Adsorbent particle size:- 0.052 mm; Temperature:-50$^{0}$C; Adsorbent amount:-6.75gm/L of model fuel, Stirrer speed:- 1000 rpm)**

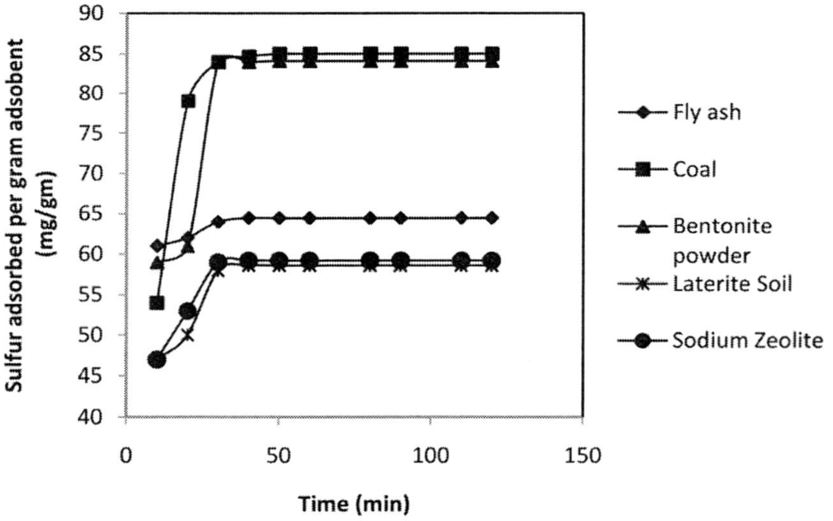


**Figure. 1D Adsorptive desulfurization of MF 4**

**(Sulfur:-900 ppm; Adsorbent particle size:- 0.052 mm; Temperature:-50$^{0}$C; Adsorbent amount:-6.75gm/L of model fuel, Stirrer speed:- 1000 rpm)**

Fig 1A,1B,1C and 1D showed the adsorptive desulfurization of MF-1, 2, 3 and 4 by using coal dust, bentonite, flyash, laterite and zeolite. The highest sulfur adsorption per gram adsorbent(mg/g) were 44.4, 29.3, 18.66 and 85 for MF 1,2,3 and 4 respectively using coal dust, 43.92, 25.9, 17.9 and 84.1 for MF 1,2,3 and 4 respectively using bentonite, 43.86, 21.8, 2.66 and 64.45 for MF 1,2,3 and 4 respectively using flyash, 43.43, 18.7, 1.93 and 58.6 for MF 1,2,3 and 4 respectively using laterite and 43.29, 18.8, 1.91 and 59.2 for MF 1,2,3 and 4 respectively using sodium zeolite. .

The adsorption result shows that coal dust is the best adsorbent and the effect of different process parameters of adsorptive desulfurization has been studied using coal dust. In this study MF 4 was taken as a fuel sample of choice. The reason for best performance of coal dust as adsorbent, compared to other four adsorbents is its high surface area which causes high mass transfer between adsorbate and adsorbent. Table 1and 2 show BET surface area and XRF elemental analysis of all adsorbents.

**Table 1.** BET surface area of adsorbents

| Adsorbent | BET surface area ($m^2/g$) |
|---|---|
| Coal dust | 149.40 |
| Laterite | 60.00 |
| Bentonite | 33.06 |
| Sodium Zeolite | 25.24 |
| Flyash | 19.25 |

**Table 2.** Elemental analysis (wt%) of all adsorbents using XRF

| Adsorbents | $SiO_2$ | $Al_2O_3$ | $SO_3$ | CaO | MgO | $Na_2O$ | $Fe_2O_3$ |
|---|---|---|---|---|---|---|---|
| Coal Dust | 42.43 | 22.169 | 9.571 | 5.099 | 0 | 0 | 13.491 |
| Bentonite | 53.28 | 18.27 | 0 | 0 | 3.027 | 7.311 | 12.975 |
| Flyash | 25.85 | 10.68 | 0 | 47.34 | 7.604 | 0 | 4.074 |
| Laterite soil | 30.551 | 17.28 | 0 | 0 | 0 | 0 | 49.37 |
| Sodium Zeolite | 41.58 | 32.96 | 0 | 0 | 0 | 23.67 | 0 |

**Parametric Study on adsorptive desulfurization of model fuel 4**

**Effect of stirrer speed variation**

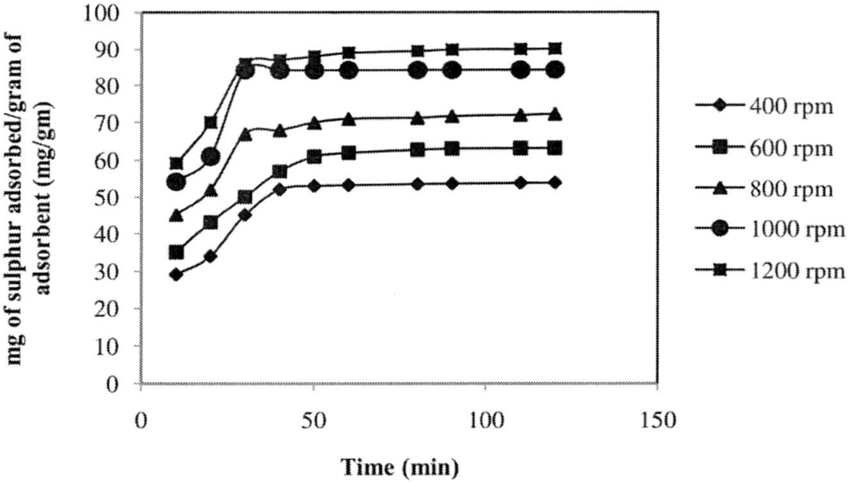


**Figure. 2 Stirrer speed variation**

**(Total Sulfur in model fuel:-900 ppm; Adsorbent particle size:- 0.052mm; Temperature:-50$^0$C; Adsorbent amount:-6.75gm/L of model fuel)**

Fig 2 shows the effect of stirrer speed on adsorption efficiency (mg/g) of coal dust. It is observed that increase in stirrer speed from 400 rpm to 1000 rpm increases the adsorption efficiency from 53.87 to 84.29. This may be due to increase in the mass transfer between adsorbate and adsorbent. But when the stirrer speed was increased from 1000 rpm to 1200 rpm the change in adsorption efficiency was not significant, which may be due to the reason that resistance to mass transfer has no prominent influence on adsorption at that range of speed [12]. So all the subsequent parametric studies were done at 1000 rpm.

**Effect of adsorbent particle size variation**

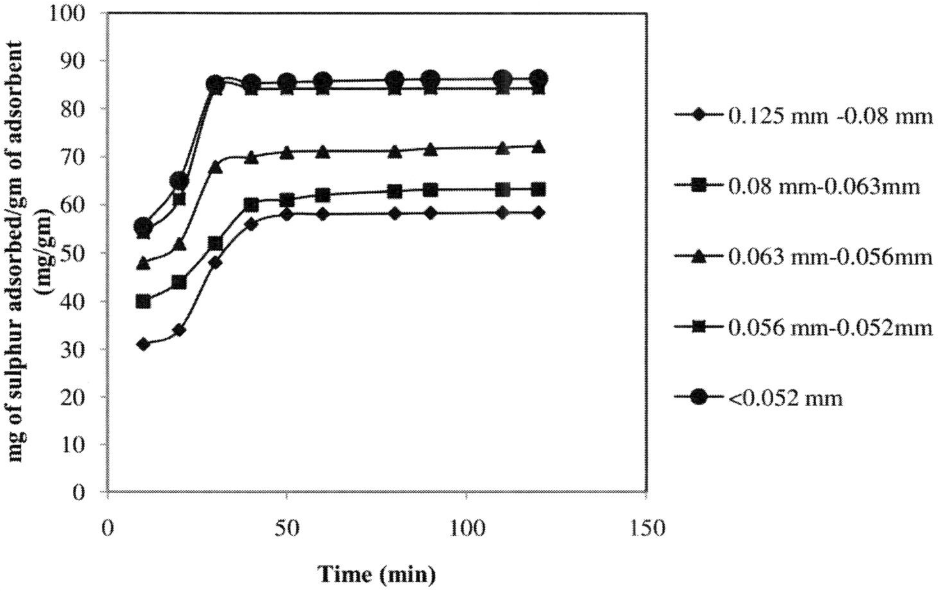


**Figure. 3 Adsorbent particle size variation**

**(Total Sulfur in model fuel:-900 ppm; Temperature:-50$^0$C; Adsorbent amount:-6.75gm/L of model fuel;Stirrer speed:-1000 rpm)**

Fig 3 shows the adsorbent particle size effect, the size ranging from 0.125 mm to 0.052 mm on its adsorption efficiency (mg/g). It has been observed from the figure that, as particle size decreases from 0.125 to 0.052 mm, the efficiency of adsorption increases. This may be because of the reason that with decrease in particle size, the surface area is expected to increase, which in turn increases the efficiency from 58.4 to 86.34 from particle size 0.125 to 0.052 mm [12]. Moreover, when the size of adsorbent is reduce to less than 0.052 mm, the adsorption capacity is not significantly increased. This may be due to the fact that, the increase in surface area after that size was not significant, hence the increase in adsorption capacity was not prominent. The rest of the experiments were performed with 0.052 mm size as internal mass transfer resistance may be negligible at this stage.

**Influence of initial sulfur concentration**

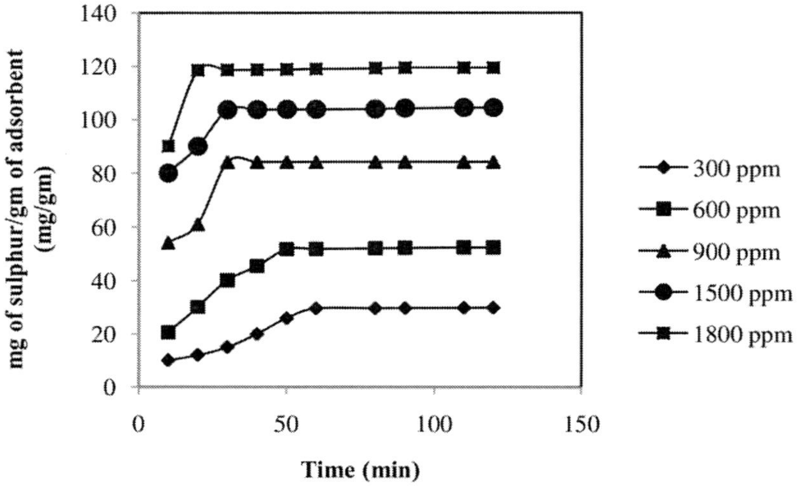


**Figure. 4 Initial sulfur concentration variations**

**(Stirrer speed:-1000 rpm; Adsorbent particle size:- 0.052mm; Temperature:-50$^0$C; Adsorbent amount:- 6.75gm/L of model fuel)**

The adsorption experiments were performed with different initial sulfur concentration of the model fuel, ranging from 300 to 1800 ppm and Fig 4 shows the effect of initial sulfur concentration on the adsorption capacity of coal. As the initial concentration of sulfur increases, intake capacity of adsorbent or adsorption rate also increases at the same time. Increase in initial concentration leads to increase in driving force (concentration gradient) for mass transfer. From the figure it is observed that, sulfur removal on the adsorption increases with time, reaches its maximum and then forms a plataue with a little decreasing trend for all the concentration. The maximum removal of sulfur reaches earlier for higher initial concentration of sulfur in model fuel after that, no improvement in adsorption with time was observed. The maximum removal of sulfur for initial concentration of 1500 and 1800 ppm obtained are 105 at 40 min and 120 mg/gm at 30 min respectively. This may be due to saturation of active sites of coal or pore blocking/active site blocking at higher sulfur concentration [1,12].

**Effect of operating temperature**

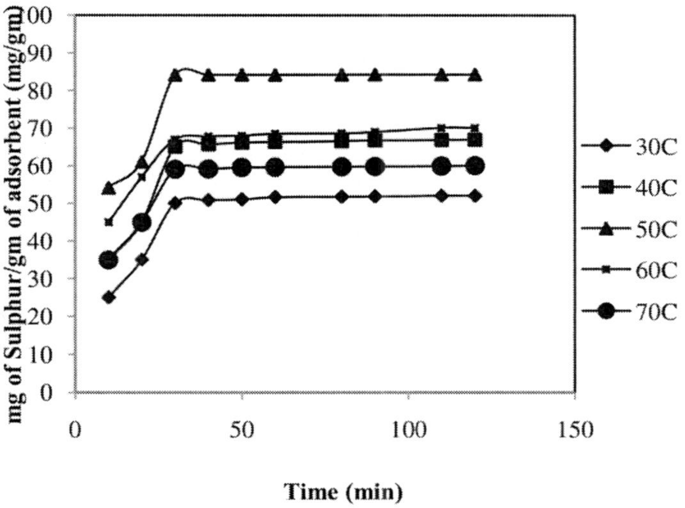


**Figure. 5 Adsorption temperature variation**

**(Stirrer speed:-1000 rpm; Adsorbent particle size:- 0.052mm; Sulfur Concentration in model fuel:-900 ppm; Adsorbent amount:-6.75gm/L of model fuel)**

Temperature for adsorption was varied from 30 to 70 $^0$C and the effect of temperature on adsorption (mg/gm) is shown in Fig 5. It has been observed from the figure that, the adsorption of sulfur increases with increase in temperature from 30 to 50 $^0$C in a regular manner, after that it decreases as the temperature increase from 50 to 70 $^0$C. This nature may be explained by the reason that, with increase in temperature, pore size of adsorbent as well as diffusion rate may be increased but above 50 $^0$C, desorption of sulfur compounds from adsorbent surface may take place at higher temperature , hence the overall adsorption decreases [1,9,10].

## Effect of variation of adsorbent amount

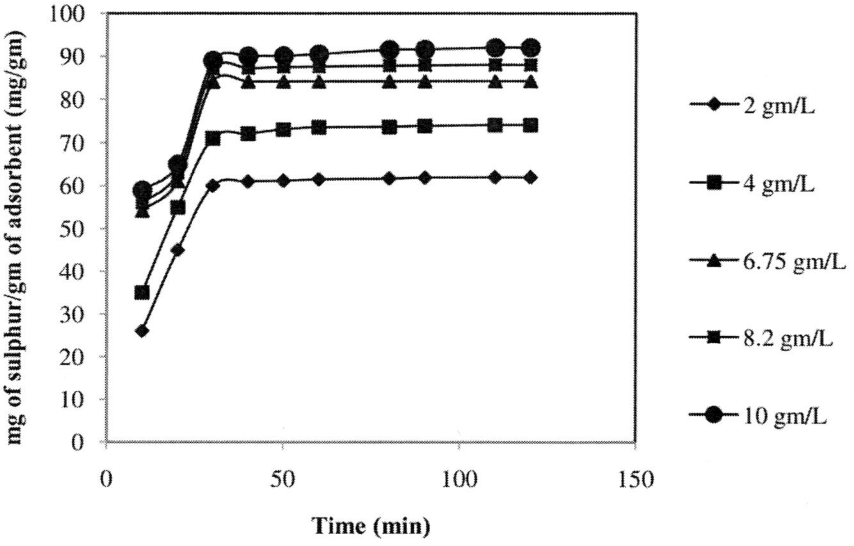


**Figure. 6 Adsorbent amount variation**

**(Stirrer speed:-1000 rpm; Adsorbent particle size:- 0.052mm; Sulfur Concentration in model fuel:-900 ppm; Temperature:-50$^0$C)**

Fig 6 shows the effect of adsorbent (coal) quantity on adsorbent capacity. It is observed from the figure that the adsorption capacity increases from 62 to 92 as the amount of adsorbent increases from 2 to 10 gm/L. This may be due to the availability of more adsorbent which provides more active sites for adsorption [1,9,10].

**Reusability of adsorbent**

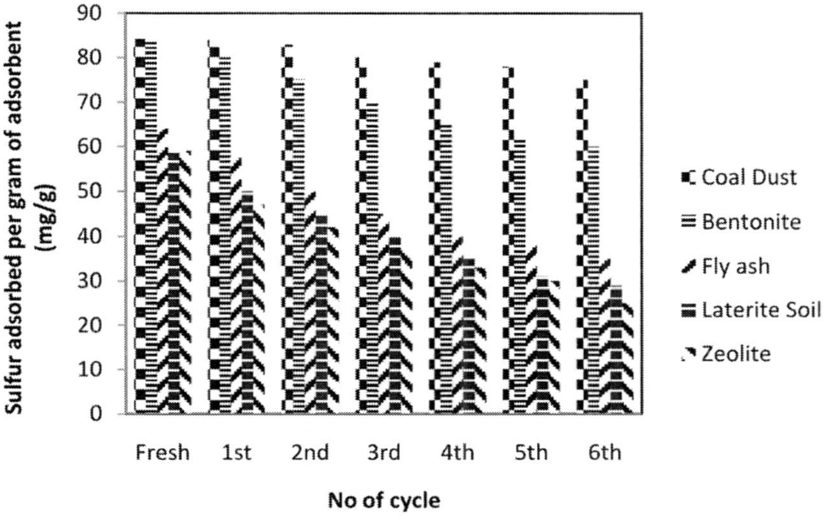


**Figure. 7 Reusability of adsorbents**

**(Sulfur:-900 ppm; Adsorbent particle size:- 0.052 mm; Temperature:-50$^0$C; Adsorbent amount:- 6.75gm/L of model fuel, Stirrer speed:- 1000 rpm)**

Fig 7 shows the reusability of all adsorbents on sulfur adsorption. It is observed from the figure that the reusability efficiency of coal dust is quite high compared the other four adsorbents, after 3$^{rd}$ cycle, the sulfur adsorption decreased only 5% for coal where as for bentonite, fly ash, laterite and zeolite the sulfur adsorption decreased 15%, 30%, 32% and 37% respectively.

### Adsorption Equilibrium Isotherm

From different adsorption equilibrium isotherms, Langmuir, Freundlich, BET, Halsay and Tempkin have been chosen to fit the experimental data and search the best fitted one. The values of all adsorption equilibrium isotherm constants for different models are showed in Table 3. It has been observed that the Langmuir adsorption isotherm is the best fitted isotherm model with highest $R^2$ value

**Table 3. Values of adsorption equilibrium isotherm constants**

| Isotherm Model | Parameters | value at 50 $^0$C | $R^2$ |
|---|---|---|---|
| 1. Langmuir | K | 0.00183 | 0.991 |
| | $Q_m$ | 200 | |
| 2. Freundlich | $K_F$ | 1.945 | 0.985 |
| | 1/n | 1.67 | |
| 3. BET | $q_s$ | 16.95 | 0.932 |
| | $C_{BET}$ | 9.83 | |
| 4. Halsay | $K_{Ha}$ | 0.321 | 0.957 |
| | $n_{Ha}$ | - 1.66 | |
| 5. Tempkin | $A_T$ | 0.0198 | 0.959 |
| | B | 38.94 | |

The experimental conditions were all constant (Stirrer speed:- 1000 rpm;Temperature:- 50$^0$C; Time:- 2hr; Adsorbent particle size:- 0.052 mm; Adsorbent amount:- 6.75 gm/L of adsorbent; Sulfur Conc:- 900 ppm)

### Conclusion

Adsorptive desulfurization of model fuels was performed by using coal dust, bentonite, fly ash, laterite and zeolite. The best result was performed by coal dust [84.22 mg sulfur adsorbed/gram of adsorbents] with a high surface area of 149.40 $m^2$/gm. There was a significant effect of process parameters such as time, temperature, initial sulfur concentration, stirrer speed, adsorbent amount and adsorbent particle size on the adsorptive desulfurization process by using best adsorbent. The Langmuir isotherm was best fitted with equilibrium isotherm data.

## Nomenclature

$q_t$ Adsorption capacity at time t (mg/g)

$q_e$ Adsorption capacity at equilibrium(mg/g)

$q_s$ Amount of sulfur adsorbed per gram of adsorbent at saturation condition or theoretical isotherm saturation capacity (mg/g)

$C_e$ Residual concentration of sulfur compound in model fuel at equilibrium. (mg/L)

$C_i$ Initial concentration of sulfur compound in model fuel (mg/L)

$C_t$ Residual concentration of sulfur compound in model fuel at time t (mg/L)

V Volume of model fuel (L)

W Mass of adsorbent (gm).

$Q_m$ Maximum monolayer coverage capacities (mg/g).

K Langmuir isotherm constant (L/mg)

$K_F$ Freundlich isotherm constant related to adsorption capacity (mg/g)

n Adsorption intensity.

$C_s$ Adsorbate monolayer saturation concentration (mg/L)

$C_{BET}$ BET adsorption isotherm relating to the energy of surface interaction (L/mg)

$K_{Ha}$, $n_{Ha}$ Halsey isotherm constants (mg/L)

$A_T$ Tempkin isotherm equilibrium binding constant (L/g)

R Universal gas constant (J/mol/K)

T Absolute temperature ($^0$K)

$b_T$ Tempkin isotherm constant

$K_D$ Distribution coefficient

## References


1. Ishaq, M., Siraj Sultan, Imtiaz Ahmad, Hameed Ullah, Muhammad Yaseen and Alia Amir (2015), " Adsorptive desulfurization of model oil using untreated acid activated and magnetite nanoparticle loaded bentonite as adsorbent," Journal of Saudi Chemical Society, http://dx.doi.org/10.1016/j.jscs.2015.02.003

2. Muzic, M., Katica S. Bionda and Zoran Gomzi (2008), "Kinetic and statistical of Adsorptive desulfurization of diesel fuel on commercial activated carbons," Chemical Engineering Technology, pp. 355-364

3. Velu, S., Xiaoliang Ma and Chunshan Song (2002), "Zeolite based adsorbents for Desulfurization of Jet fuel by selective adsorption," Fuel Chemistry Division Preprints, pp. 447-448

4. Palomino, Jessica M., Dat T. Tran, Jesse L. Hauser, Hong Dong and Scott R.J. Oliver (2014), "Mesoporous silica nanoparticles for high capacity Adsorptive Desulfurization," Journal of Materials Chemistry A, 14890-14895

5. Dasgupta, S., Pushpa Gupta, Aarti, Anshu Nanoti, Amar N. Goswami, Madhukar O. Garg, Elisabeth Tangstad, Ornulv B. Vistad, Arne Karlsson and Michael Stocker (2013), "Adsorptive desulfurization of diesel by regenerable nickel based adsorbents," Fuel, pp. 184-189

6. Xu, X., Shuyang Zhang, Peiwen Li and Yuesong Shen (2013), "Equilibrium and kinetics of Jet-A fuel desulfurization by selective adsorption at room temperature," Fuel, pp. 172-179.

7. Adekanmi, Abel A. and Aberuagba Folorunsho (2012), "Comparative analysis of adsorptive desulfurization of crude oil by Manganese di Oxide and Zinc Oxide," Research Journal of Chemical Science, pp. 14-20

8. Shi, Y., Xiangwen Zhang and Guozhu Liu (2015), "Adsorptive desulfurization performances of ordered mesoporous Carbon with tailored textural and surface properties," Fuel, pp. 565-571

9. Ahmad, W., Imtiaz Ahmad, Muhammad Ishaq and Khudiji Ihsan (2014), "Adsorptive desulfurization of kerosene and diesel oil by Zn impregnated montmorollonite clay," Arabian Journal of Chemistry, http://dx.doi.org/10.1016/j.arabjc.2013.12.025

10. Khodadadi, A., Torabi M. Angaji, Talebizadeh A. Rafsanjani and A. Yonesi (2012), "Adsorptive desulfurization of Diesel fuel with nano Copper Oxide," 4th International Conference on Nanostructures, 2012, 1197-1199.

11. Kumar, S., Vimal C. Srivastava and R. P. Badoni (2011), "Studies on Adsorptive desulfurization by Zirconia based adsorbents," Fuel, 3209-3216.

12. Patil, S M., Y.C. Bhattacharyulu and S.R. Kulkarni (2011), "Desulfurization of hydrocarbon liquid fuels by adsorption ," Journal of Engineering Research Studies, pp. 81-98

# 513254 Oil to Chemicals: Technological Approaches and Advanced Process Configurations to Convert Light Crude Oils to Olefins

**Abstract:**

Interest in the topic of crude oil conversion direct to chemicals is not new, but it has gained recent interest due to new drivers â€" first, the advancements in technology to produce chemicals including the increased versatility of cracking and unconventional olefin and BTX swing processes, second, the prospect of lower crude pricing for longer making it an attractive feedstock for chemical producers, and third, the recent surplus supply of light crude oils. Two recent public announcements in this area include ExxonMobilâ€™s modified steam cracker in Singapore and the joint venture project between Saudi Aramco and SABIC.

This presentation will highlight the causes for increased interest in this area, including the imbalance in projected growth between transportation fossil-fuels and petrochemicals, as well as highlight and summarize the technologies that are being developed to make these conversion processes a reality. The presentation will also illustrate why future commercial crude oil to chemical installations will likely be installed in a modular fashion and greatly benefit chemical producers.

# 512889 Experimental and Numerical Study on the Effects of Spiral Guide Vane Turns on Cyclone Performance

**Abstract:**

Cyclone separators are utilized in a variety of industries to separate solid particles. They are popular because they are low in cost, easily maintained, easily operated, free of moving components, and highly reliable even under extreme working conditions. There are many geometric parameters that impact cyclone performance, but collection efficiency and pressure drop are particularly influential. The inlet geometry is also a crucial parameter, as it controls the initial spiral flows which dominate the flow field inside the cyclone.

This paper presents an experimental and numerical study on a tangential-inlet cyclone separator with different numbers of spiral guide vane turns. Numerical pressure drop results were in close agreement with the experimental data. The spiral guide vane was also found to considerably influence the velocity distribution, turbulence intensity, pressure drop and collection efficiency in the cyclone. A critical value of spiral guide vane turns appeared below or above which there was a marked increase in separation efficiency, pressure drop, and tangential velocity. Compared to a cyclone with zero spiral guide vane turns, the maximal decrease in separation efficiency in the cyclone with the critical spiral guide vane turn value (one turn) was 2%. The maximum-efficiency inlet velocity appeared to exist independent of spiral guide vane turns, as inlet velocity affects the radial distance traveled by the rebounded particles from the inner wall. The results presented here may provide a workable reference for the effects of spiral guide vanes on the flow field and corresponding performance in cyclone separators.

# 512275 Continuous Non Centrifugal Phase Separation in Biphasic Whole-Cell Biocatalysis Using a Lab-Scale Prototype

**Abstract:**

Biphasic whole-cell bio-processes have already reached levels of industrial relevance and are known to be efficient and green alternatives to common chemical synthesis routes. The main bottleneck for an industrial implementation of these high-potential (high stereo selectivity, high product titers) biphasic whole-cell bio-processes is the presence of particles (cells) resulting in the formation of stable Pickering type emulsions hindering cost-efficient and effective DSP.

In contrast to state of the art DSP-processes e.g. centrifugation or the use of de-emulsifiers, the phenomenon of catastrophic phase inversion (CPI) applied within this work, easily achieves phase separation by addition of dispersed phase. Due to a sudden switch of emulsion type, Pickering type emulsions are destabilized efficiently and phase separation is achieved with minimal effort.

Based on a patent filed at TU Dortmund, a process concept (termed: Applied Catastrophic Phase Inversion; ACPI) was developed, enabling continuous phase separation of stable emulsion by using the CPI phenomenon. A fully automated pilot-scale prototype was planned and constructed. Various sensors e.g. for conductivity or level-control allow for process monitoring and control. First runs were performed for defining a process window (necessary amount of disperse phase, flow and mixing rates, settler setup, etc.). In further runs also the stability of the process was evaluated by introducing external disturbances. Especially the influence of varying O/W ratios in the feed emulsion) were analyzed. All results will be presented, demonstrating the efficiency and applicability of the concept for phase separation of stable Pickering type emulsions.

ACPI benefits from its simple setup, low operating costs, a broad applicability irrespective of product concentration and success of the initial biotransformation. This fact qualifies ACPI as a considerable alternative as first separation step in the DSP of biphasic whole-cell biotransformations.

# 509914 Large-Scale Separation Efficiency of Dispersed Oil Micro-Droplets from Aqueous Phase with Magnetic Nanoparticles

**Abstract:**

Introduction: Gravity separators are favored with regard to removing dispersed oil droplets from water, due to their low operating and maintenance cost. This conventional solution takes advantage of gravity or density differences in oil and water. The separation process usually takes 12-24 hours and oil-removal efficiency will reach 69-82%. With the elevated environmental concerns about the produced formation fluids, efficient and fast methods are required to process large volumes of water at water treatment facilities.

Theory: Application of magnetic nanoparticles (MNPs) in removing oil from oil-in-water emulsion is proved efficient at lab-scale. The term magnetophoresis (MAP) refers to the phenomenon of a magnetic particle moving through a viscous fluid medium in an external non-uniform magnetic field. An efficiently designed MAP method successfully removes 95-99.5% of oil colloids in less than one hour. MAP theory is favorably applied for particle separation, especially cell separation in biomedical and material sciences. MNPs exhibit superparamagnetic properties and respond to an external magnetic field easily until the field is removed. This allows the mechanical control of MNPs motion in a carrier fluid by magnetic field, which serves as the fundamental concept for a MNP-based separator design.

Methods: A comprehensive study of MAP involves nanoparticle synthesis, advanced magnetism, particle-laden flow dynamics, particle tracing, and analytical and numerical analyses. This study presents numerical modeling of magnetic nanoparticle motion in a magnetic fluid medium under an external magnetic field. Strong emphasis is given to the compilation of the available experimental data and matching the observed experimental results. A comprehensive semi-analytic diagnostic model is presented to understand the fundamental dynamic of the key design parameters involved, followed by a more complex, tuned numerical model.

Contributions: The contributions of this study are (1) to identify the feasibility of magnetic separation in oil removal, and (2) to optimize a magnetic separator design by investigating optimum MNP concentration, MNP diameter, and required external magnetic field strength, and (3) to quantify the optimum MNP separation efficiency when large-scale flowrates are applied.

# 512897 Experimental Analysis of the Tangential Velocity Fluctuations in the Cyclone

**Abstract:**

In order to get further insight into the characteristics of swirling flow in the gas flow field in the cyclone,The tangential velocity fluctuation characteristics of swirling flow in the cyclone were studied based on measured by Hot Wire Anemometry (HWA). The results show that the real tangential velocity fluctuates obviously in the center region of the cyclone, however, near the wall of the cyclone, the real tangential velocity has a little fluctuation. Furthermore, the real velocity has a basic frequency in the cyclone, which is mainly depend on the tangential velocity in the vortex. Also the amplitude in the center region is bigger than that near the wall. The real tangential velocity fluctuation indicates that there exists an instability of swirling flow in the cyclone, which results in the rotation center of the swirling flow deviating from the geometric center of the cyclone and produces the quasi-periodic oscillation of swirling flow.

# 513242 Development of a Novel Nanofiltration Membrane Based on the Polydopamine Chemistry

**Abstract:**

This work reports on the fabrication of new nanofiltration (NF) membrane with a robust and permeable Novel methodology for facilely fabricating nanofiltration (NF) composite membrane has been successfully developed by employing nucleophilic nature of polydopamine (PDA) chemistry. The self-polymerized PDA coating over polysulfone (PSf) substrate was utilized as a key intermediate layer for trimesoyl chloride (TMC) grafting followed by poly(ethyleneimine) (PEI) deposition to construct the hierarchically structured separation layer of NF membrane. In contrast to the electrophilic quinone moieties of PDA layer usually involved in the Michael addition and Schiff base reactions with polymeric amines for membrane preparation in previous reports, the phenolic hydroxyl groups of catechol moieties as well as amine groups at the PDA layer possess nucleophilic nature, which are capable of quickly coupling with the highly reactive acyl chlorides to form ester and amide bonds in the step of TMC grafting, resulting in TMC moieties covalently anchored at the PDA layer with free acyl chloride groups. Such created acyl chlorides are further coupled with the amine groups of branched PEI polymer in the PEI deposition procedure to form the stable amide bonds linking the PDA base layer and PEI upper layer in the resulting hierarchical separation layer. The properties of membranes prepared at different stages were characterized with respect to surface chemistry, pore properties, and separation performances to understand deeply the newly developed methodology for membrane preparation. Further studies focusing on NF properties and stability of the developed NF membrane revealed that such membrane shows high efficiencies in retention of divalent cations, small organic molecules as well as heavy metal ions, and exhibits desirable thermal stability and long-term performance stability.

The resulting PEI deposited membrane M-PEI shows NF characteristic with mean pore size of 0.23 nm and MWCO of 292 Da, while possesses high rejection efficiency towards divalent cation Mg2+, small organic solutes, and heavy metal ions because of its positively charged surface feature and small pore size. Surface chemistry studies reveal that M-PEI has very smooth and hydrophilic layer surface owing to the existence of uniform PEI coverage with plenty of hydrophilic free amine groups. Furthermore, the M-PEI NF membrane is proved to be thermally stable in the range of feed temperature tested, owing to the strong adhesion of PDA base layer on the PSf substrate, as well as the existence of covalent bonds linking the middle TMC grafting layer with both PDA coating and PEI upper layer. This work demonstrates that the composite membrane with high separation efficiency could be effectively fabricated on the basis of nucleophilic nature of polydopamine. Further study focusing on the functionalities such as chlorine resistance and anti-fouling of the desalination membrane fabricated on basis of this newly developed methodology is ongoing in our lab.

# 513651 Modeling Hydrocyclones Using Two Fluid Model (Eulerian-Eulerian) Approach

**Abstract:**

Hydrocyclones are vastly used in coal separation, petrochemical engineering, mineral processing and pulping. Tangential injection of the liquid into a cylindrical chamber causes the development of a vortex. The chamber has a restricted axial bottom outlet such that all of the liquid in the vortex cannot escape via this outlet. A portion of the liquid has to reverse its path and flow towards the axial top outlet. This reverse flow continues to rotate and an air core develops due to lower pressure at the axis of rotation. As a result the heavy components move circumferentially toward the wall of the vessel where they agglomerate and spin down the wall to the outlet at the bottom of the vessel. Light components move toward the axis of the hydrocyclone where they again move up toward the overflow outlet at the top of the vessel.

Numerical simulation is a powerful tool to understand the flow behavior, calculate the pressure drop, particle trajectories and collection efficiency of hydrocyclones. Also it can reduce the experimental costs required for testing and optimization. While extensive work has been done on performance, scale-up, and simulation of hydrocyclones the effect of used approach in the results is not yet clear and there are discrepancies between models according to mesh type, turbulence model and seeding rate.

In this work, the Two Fluid Model (TFM) approach will be compared with Lagrangian Particle Tracking (LPT) approach. In case of low feeding solids, the CFD-LPT approach results are accurate and computationally efficient. However, as the effect of inter-particle interactions as well as the reaction of particles on the fluid are ignored its application is limited. On the other hand, in TFM both the fluid and solid phases are treated as interpenetrating continuum media and therefore it can handle feeding solids at different concentrations. We used Reynolds Stress Model and compared various cases to show the advantages of TFM approach. In addition, we showed that an increase in the inlet pressure will increase the static pressure differential along the radius within the cyclone body hence more water split into overflow. At the end, we discuss the effect of particle density on efficiency of cyclone.

# Separation Performance of Supersonic Separator with a Rear-Set Helical Guide Blade

*Huirong Liang[1,2], Shuai Zhang[2], Yong Kang[2], Kegang Ling[1]*
*(1) Petroleum Engineering Department, University of North Dakota, Grand Forks, ND,*
*(2) School of Mechanical Engineering, Xi'an Shiyou University, Xi'an, China*

**Abstract:** The supersonic separation is a new approach to dehydrate the natural gas in recent years. In the conventional supersonic separator, a cyclone or a delta wing is employed to generate the swirling flow. Unfortunately, the separation efficiency of conventional supersonic separator is low. In order to overcome the disadvantages, helical guide blade was designed to incorporate into the swirling device which is fixed in the middle of the separation straight tube in this paper. The computational fluid dynamics modeling was performed to investigate the flow characteristics in the supersonic separator and optimize the structure of the blade. The swirling flow generated from the blade promoted the separation of the droplets from the gas. The optimization results showed that a pitch of 200 mm and a single blade are the best geometry structure for the swirling device in the supersonic separator in the dehydration process, which can create the stable flow field and benefit the separation effect.

## 1 Introduction

The natural gas collected from the gas wells is generally saturated with large volumes of water vapor, which inevitably causes clogging and corrosion in natural gas transportation pipelines, valves, and other auxiliary equipment when water vapor condenses out. It also seriously affects the heating value and quality of the natural gas, and reduces the utilization efficiency of the natural gas. Therefore, it is imperative to effectively remove the free water or saturated water in the natural gas treatment process. There are various methods for the liquid separation from the gas[1-3]. Supersonic separator is a new type of gas-liquid separation equipment with the advantages of low energy consumption, safe and efficient operation, and pollution-free to environment, which has attracted the attention in the separation sector[4-6]. Jassim et al.[7] and Karimi and Abdi[8] researched the single-phase flow of natural gas in the de Laval nozzle under a high pressure condition. Yang et al.[9-10] investigated the characteristics of the pressure recovery and real gas using CFD method. Malyshkina[11] studied the effect of the Mach number on the liquid component and obtained the optimum Mach number of 1.4-1.8 for a supersonic separator. Wen et al.[12] optimized the geometry structure of the diffuser in the supersonic separator under the conditions of single phase flow.

In a conventional supersonic separator, the shockwave usually occurs at the expanding section of the de Laval nozzle, which results in a short low-temperature section and a high flow resistance, consequently a low separation efficiency[13-15]. In order to improve the separation performance of the supersonic separator, a new type of the swirling device was designed. The flow characteristics were then investigated in our newly designed supersonic low-temperature gas-liquid separator using three-dimensional numerical simulation. And the effects of the structure of the helical guide blade on the separation efficiency were also researched in this paper.

## 2 Supersonic separators

The conventional supersonic gas-liquid separator is mainly composed of a de Laval nozzle, a separation straight tube, an annular drainage tube and a diffuser as illustrated in Fig.1. The de Laval nozzle is a key part including a converging tube, a throat tube, and a diverging tube. In the conventional structure, the straight tube is usually integrated with a cyclone to generate a strong vortex flow. Because of the strongly negative effect of the shockwave, the separation efficiency is relatively low. To overcome this disadvantage, instead of the cyclone a swirling device, which uses a helical guide blade, was designed and installed in the middle of the separation straight tube in the separator.

The design of the supersonic separator is based on the theories of supersonic wind tunnels, thermodynamics, and fluid mechanics. The saturated natural gas experiences an adiabatic expansion when flowing through a special variable-section tube of the de Laval nozzle. The temperature drops rapidly to the dew point of the gas and the pressure energy can be converted into kinetic energy. With that a flow field forms with low temperature, low pressure, and high velocity. The liquid droplets start to condense in the diverging section of the de Laval nozzle and are centrifuged onto the wall when the mixture of liquid and gas flows through the swirling device. They are collected and drained through the annular drainage tube. Finally, the dry gas exits through the diffuser.

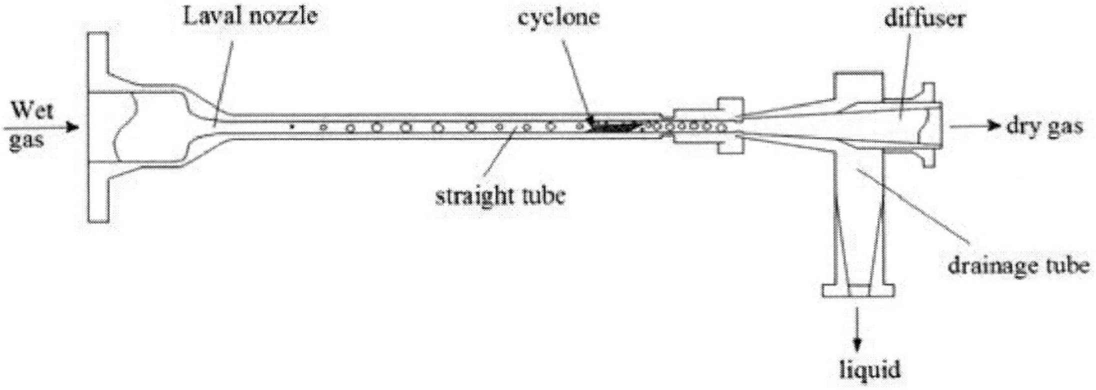


**Fig 1. Structure schematic diagram of the conventional supersonic separator**

## 3 Calculation model

### *3.1 Conservation equations*

The internal flow in the supersonic separator follows the laws of the fluid dynamics, i.e. three governing equations, including the continuity, momentum and energy equations[14]. Because natural gas is compressible, the energy conversion of the gas flow is involved.

Continuity equation:

$$\frac{\partial \rho}{\partial t} + div(\rho \vec{u}) = 0$$

Momentum equation:

$$\frac{\partial \rho u}{\partial t} + div(\rho u \vec{u}) = -\frac{\partial p}{\partial x} + \frac{\partial \tau_{xx}}{\partial x} + \frac{\partial \tau_{yx}}{\partial y} + \frac{\partial \tau_{zx}}{\partial z} + F_x$$

$$\frac{\partial \rho \upsilon}{\partial t} + div(\rho \upsilon \vec{u}) = -\frac{\partial p}{\partial y} + \frac{\partial \tau_{xy}}{\partial x} + \frac{\partial \tau_{yy}}{\partial y} + \frac{\partial \tau_{zy}}{\partial z} + F_y$$

$$\frac{\partial \rho \omega}{\partial t} + div(\rho \omega \vec{u}) = -\frac{\partial p}{\partial z} + \frac{\partial \tau_{xz}}{\partial x} + \frac{\partial \tau_{yz}}{\partial y} + \frac{\partial \tau_{zz}}{\partial z} + F_z$$

Energy equation:

$$\frac{\partial(\rho T)}{\partial t} + div(\rho \vec{u} T) = div\left[\frac{\kappa}{c_p} \cdot gradT\right] + S_T$$

Where $\rho$ is the density of the gas; $t$ is the time; $\vec{u}$ is the velocity vector; $p$ is the pressure; $T$ is the temperature; $F_x$, $F_y$ and $F_z$ are the volume force on the microelements; $\tau_{xx}$, $\tau_{xy}$ and $\tau_{xz}$ are the component of the viscous stress acting on the surface of the microelements.

*3.2 Turbulence model*

In the supersonic separator, the flow after the swirling device is strongly turbulent so that the anisotropic effect is very significant especially when the gas flow occurs at supersonic speed. Therefore, it is very important to select an appropriate model to simulate the turbulent flow. According to the flow characteristics of the supersonic separator, RNG $k$-ε model is used to predict the complex swirling flow in this paper. RNG $k$-ε model involves turbulent vortices and improves the accuracy of the simulation for the swirling flow. It can satisfy the requirements for the calculation accuracy and speed.

*3.3 Numerical Schemes*

In this study, a three-dimensional calculation model of supersonic separation was established as Fig.2. The flow field is discretized using different structure grids in different sections. The finite difference method is utilized to solve the governing equations. The second-order upwind scheme is adopted to discretize the flow equations and the appropriate boundary conditions are set in the model. The pressure boundary conditions are specified at inlet and outlet of the supersonic separator, respectively. The walls of the flow field and the surface of the blade are all set as the solid walls, and moreover, no-slip and adiabatic boundaries. The standard wall function method is used to deal with the boundary turbulence.

**Fig 2. Schematic diagram of the calculation model**

## 4 Results and discussion

*4.1 The characteristics of the flow field*

In this model, the swirling device is installed in the middle of the separation straight tube, which contains a single blade with a pitch size of 150 mm. The cloud map distributions of temperature, velocity and pressure are shown in Fig.3.

From the outlet of the nozzle to the spiral blade region, the fluid keeps at a low temperature. The temperature field downstream of the spiral blade is evenly distributed. The minimum temperature is about 190 K. The temperature in the local region is slightly higher. The axial flow is smoothly transitioned to the spiral flow. The highest flow rate that can be achieved at the outlet of the nozzle is 644 m/s. The pressure cloud map shows that the pressure drop at the outlet of the nozzle is 0.36 Mpa. The loss of pressure can be converted into kinetic energy, and the temperature is reduced to the utmost. This shows that the pressure and the temperature changes in an inverse relationship to the flow rate, which is consistent with the relevant theory.

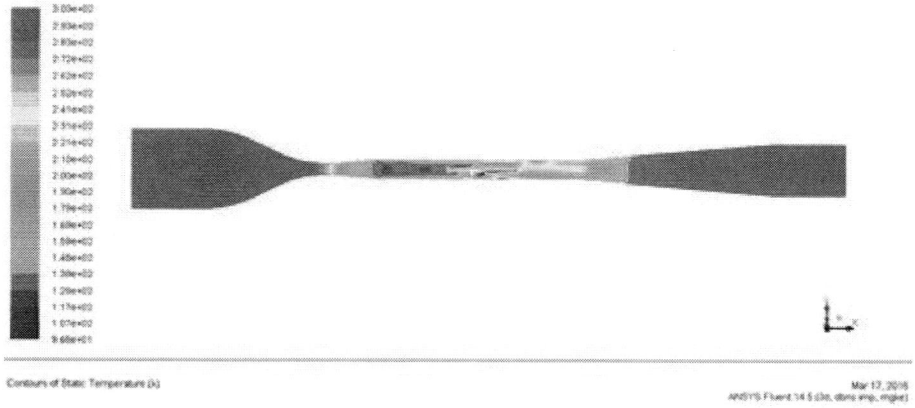

**(a) Temperature**

**(b) Velocity**

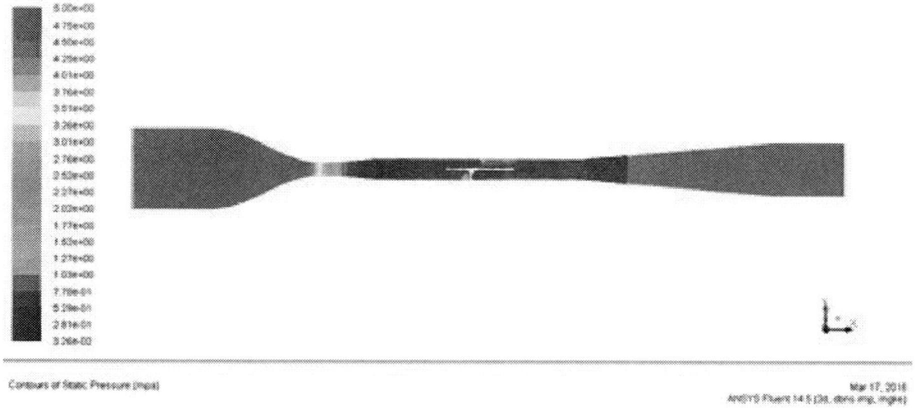

**(c) Pressure**

**Fig 3. The distribution of (a) Temperature, (b) Velocity and (c) Pressure along the axis (z=0)**

*4.2 The effect of the geometrical parameter of the guide blade*

Two parameters including the number of the blade and the pitch were considered in order to research the effect of the guide blade on the separation performance, as shown in Fig.4.

The effects of the number of the blade on the temperature, pressure and velocity are shown in Fig.4 (a), (b) and (c). It can be clearly seen from the figures that all the models reach the lowest temperature (about 140 K) at the exit of the de Laval nozzle, and the flow velocity increases to the supersonic speed (around 600 m/s). Near the leading edge of the guide blade, the temperature and pressure of the fluid suddenly rise, and the flow velocity suddenly drops, indicating that a shock wave occurs here, which is consistent with the characteristics of the flow around the obstacle. With the increase of the number of blades, the shock wave makes the fluid temperature and pressure rise faster and higher, and makes the flow velocity drops more seriously, indicating that the shock intensity increases as the number of blades increases. The stronger the shock wave intensity at the leading edge of the guide blade, the more unfavorable the stability of the flow field and the effectiveness of subsequent separation.

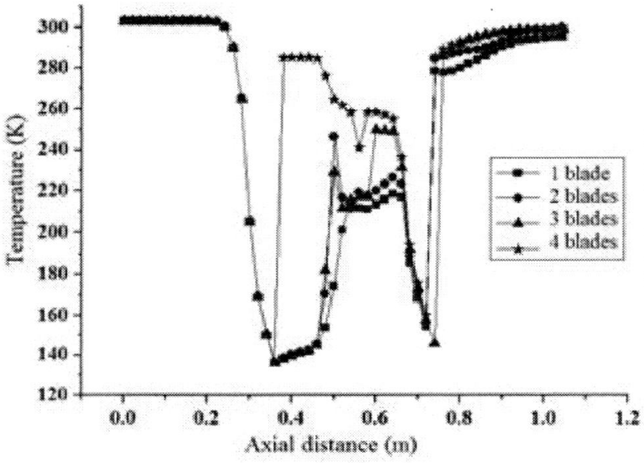


(a)

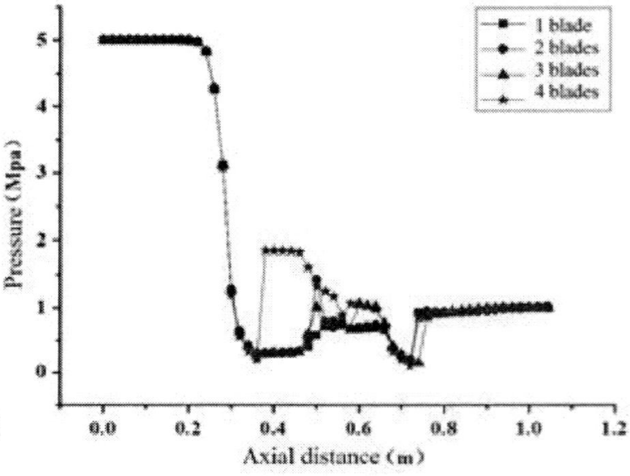


(b)

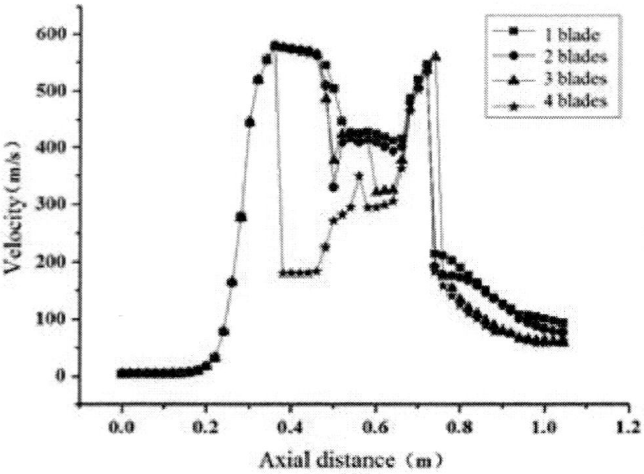


(c)

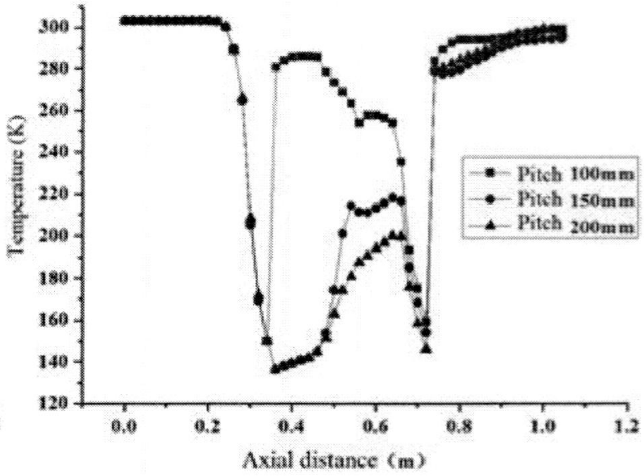


(d)

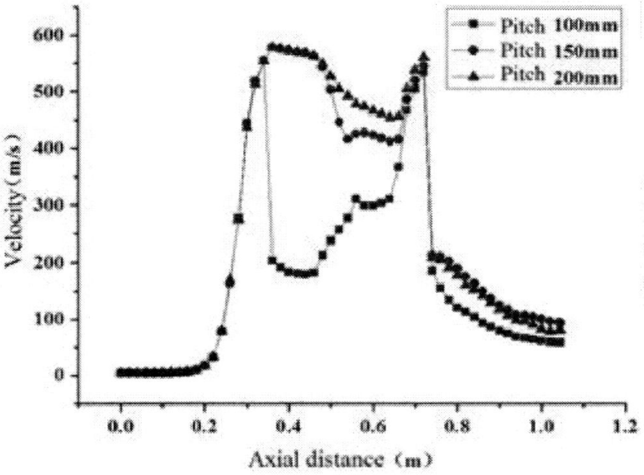


(e)

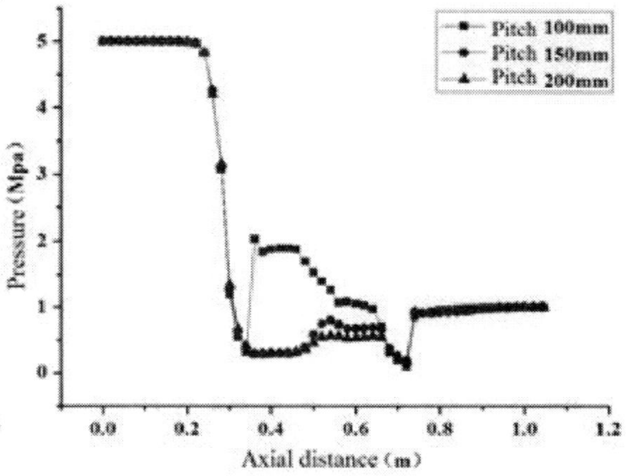


(f)

**Fig 4. The effects of the pitch or the blade number on Temperature, Pressure and Velocity**

The comparison curve shows that with the increase of the number of blades, the shock wave occurs to get close to the throat (x=0.29 m). The shock wave of the four-bladed model is closely next to the throat, where the temperature rises to 285 K, the flow velocity drops to 180 m/s, and the pressure is 1.8 Mpa. The low-temperature and low-pressure supersonic flow field cannot be realized, so the droplet nucleation and growth are failed. The separation performance is therefore the worst for the four-blade model among all the models. Compared among the four models, the single-blade model achieves a low-temperature and low-pressure supersonic flow field in the nozzle and the shock wave generated by the single blade is the weakest. The temperature in the zone of the swirl separation at the downstream of the blade (x=0.56 m ~ x=0.66 m) can be maintained at 210 K with a flow rate of about 445 m/s for the single-blade model, which provides a good environment for droplet growth and cyclone separation and consequently results in a better separation performance.

Fig. 4 (c), (d) and (e) show the comparison of temperature, velocity, and pressure along the axial direction for three models with different pitch of 100 mm, 150 mm, and 200 mm, respectively. From the comparison, it can be seen that the temperature of the models with the pitch of 150 mm and 200 mm drops below 140 K in the de Laval nozzle, while the minimum temperature of the model with 100 mm pitch is around 150 K. As the pitch decreases, the expansion wave and shock wave upstream of the guide blade start to move forward to downstream, and for the model with a pitch of 100 mm, the expansion wave is close to the throat. The appearance of expansion wave and shock wave will cause the temperature and pressure to suddenly increase, and the flow velocity to suddenly drop. The model with a pitch of 100 mm had a temperature rise to 280.9 K near the throat, the flow rate decrease to subsonic (204.2 m/s), and the pressure rise to 2.03 Mpa, which cannot satisfy the required supersonic low-temperature and low-pressure environment and cannot reach the conditions for the condensation of droplets and subsequent cyclonic separation, so the model with a pitch of 100 mm has the worst separation efficiency.

The comparative curves show that the model with a pitch of 200 mm can reach the lowest temperature of 136.2 K and the maximum flow velocity of 579.6 m/s in the de Laval nozzle, which satisfies the objective consistent with the theory of supersonic dehydration devices in the low-temperature, low-pressure and high-velocity flows. At the same time, the expansion wave and shock wave are the weakest from the outlet of the nozzle to the leading edge of the blade, which has the least influence on the performance of the flow field. Wet air flow can achieve good nucleation and growth of the droplet in the nozzle. The weak shock wave intensity in the guide blade region is beneficial to the stability of the flow field and the maintenance of the low-temperature and high-speed flow environment. Although the model with a pitch of 200 mm has an increased temperature in the blade region (to 200 K) and a gradually decreased axial flow velocity (454.3 m/s), the overall change is the most gentle, still maintain the flow field of high velocity at low temperature, reduce the probability of secondary evaporation of droplets, and simultaneously ensure the continuity of the cyclone separation.

## 5 Conclusions

The guide blade is fixed in the middle of the separation straight tube downstream of the Laval nozzle in this paper. The physical model has been established and analyzed by CFD method. The fluid keeps at a low temperature from the outlet of the nozzle to the spiral blade region, and the temperature is uniformly distributed in the downstream of the spiral blade. The axial flow is smoothly transitioned to the spiral flow. At the outlet of the nozzle the highest flow rate is 644 m/s and the pressure drop is 0.36 Mpa.

The number of the blade and the pitch are closely related to the flow performance of the spiral guide blade. With the increase of the number of spiral blades, the area where the fluid collides with the blade increases, and not only the energy loss increases, but also the location of the shock wave in the swirl tube is closer to the throat, and the intensity of the shock wave is also intensified. The research results show that the model with a single blade has a best separation performance. The effect of pitch on the separation performance was analyzed. It was found that the model with a pitch of 200 mm can control the position of the shock wave well, and at the same time weakened the shock wave strength in the swirl tube, and can maintain a long low-temperature, low-pressure and high-speed flow region, which is conducive to the efficient condensation of droplets and the cyclone separation. So it demonstrated that the model with a pitch of 200 mm has a best separation performance.

## Acknowledgement

The authors thank the Science and Technology Department of Shaanxi Province for providing the financial support of the Shaanxi Province Science and Technology Co-ordination Innovation Project (2012KTCG01-12).

## References

1. I. Folgueira, I. Teijido, A. García-Abuín, D. Gómez-Díaz, A. Rumbo, Carbon dioxide absorption behavior in 2-(ethylamino) ethanol aqueous solutions, Fuel Process.Technol. 131 (2015) 14–20.
2. N. Álvarez-Gutiérrez, M.V. Gil, F. Rubiera, C. Pevida, Adsorption performance indicators for the CO2/CH4 separation: application to biomass-based activated carbons, Fuel Process. Technol. 142 (2016) 361–369.

3. G.Q. Chen, S. Kanehashi, C.M. Doherty, A.J. Hill, S.E. Kentish,Water vapor permeation through cellulose acetate membranes and its impact upon membrane separation performance for natural gas purification, J. Membr. Sci. 487 (2015) 249–255.
4. P.B. Machado, J.G. Monteiro, J.L. Medeiros, et al., Supersonic separation in onshore natural gas dew point plant, J. Nat. Gas Sci. Eng. 6 (2012) 43–49.
5. V.I. Alferov, L.A. Baguirov, L. Dmitriev, V. Feygin, S. Imaev, J.R. Lace, Supersonic nozzle efficiently separates natural gas components, Oil Gas J 103 (2005) 53–58.
6. C. Wen, X. Cao, Y. Yang, W. Li, An unconventional supersonic liquefied technology for natural gas, Energy Educ. Sci. Technol. Part A Energy Sci. Res 30 (1) (2012) 651–660.
7. E. Jassim, M.A. Abdi, Y. Muzychka, Computational fluid dynamics study for flow of natural gas through high-pressure supersonic nozzles: part 1. Real gas effects and shockwave, Pet. Sci. Technol. 26 (2008) 1757–1772.
8. A. Karimi,M.A. Abdi, Selective dehydration of high-pressure natural gas using supersonic nozzles, Chem. Eng. Process. 48 (2009) 560–568.
9. Y. Yang, C. Wen, S. Wang, Y. Feng, Theoretical and numerical analysis on pressure recovery of supersonic separators for natural gas dehydration, Appl. Energy 132 (2014) 248–253.
10. Y. Yang, C.Wen, S.Wang, Y. Feng, Numerical simulation of real gas flows in natural gas supersonic separation processing, J. Nat. Gas Sci. Eng. 21 (2014) 829–836.
11. M. Malyshkina, The procedure for investigation of the efficiency of purification of natural gases in a supersonic separator, High Temp. 48 (2010) 244–250.
12. C.Wen, X. Cao, Y. Yang,W. Li, Numerical simulation of natural gas flows in diffusers for supersonic separators, Energy 37 (2012) 195–200.
13. W.M. Jiang, Z.L. Liu, H.W. Liu, J. Zhang, X.J. Zhang, Y.X. Feng, Two dimensional simulation and analysis of the flow in supersonic separator, J. Eng. Therm. 29 (2008) 2119–2121.
14. X.W. Liu, Z.L. Liu, Y.X. Li, Numerical study of the high speed compressible flow with non-equilibrium condensation in a supersonic separator, J.Clean Energy Technol. 3 (2015) 360–366.
15. X.W. Liu, Z.L. Liu, H.Q. Wu, Flow behavior analysis of the cyclone back-placed supersonic separator, J. B. Univ. Technol. 40 (2014) 1394-1401

## 513875 Highly Selective MOF Based MMM Integrated with CNTs for $CO_2$ Separation Purpose

**Abstract:**

Carbon Nanotubes (CNTs) were synergistically integrated to polysulfone (PSF) to develop a highly selective mixed-matrix membranes (MMMs) for $CO_2$ separation purpose in a presence of a metal organic framework (MOFs). This study aims to integrate CNTs with metal organic frameworks into a MMM to separate $CO_2$ efficiently. The results show that filled MMMs with about 5 wt % CNTs showed an optimum separation performance by providing a $CO_2/N_2$ selectivity of about 50. Increasing CNTs additives resulted in in lowering the selectivity.

# AUTHOR INDEX

# AUTHOR INDEX